Three week loan

Please return on or before the last
date stamped below.
Charges are made for late return.

RUBBERLIKE ELASTICITY
A MOLECULAR PRIMER

RUBBERLIKE ELASTICITY
A MOLECULAR PRIMER

JAMES E. MARK
Department of Chemistry
University of Cincinnati
Cincinnati, Ohio

BURAK ERMAN
School of Engineering
Bogazici University
Istanbul, Turkey

WILEY

A WILEY-INTERSCIENCE PUBLICATION
JOHN WILEY & SONS
New York • Chichester • Brisbane • Toronto • Singapore

Copyright © 1988 by John Wiley & Sons, Inc.

Library of Congress Cataloging-in-Publication Data

Mark, James E., 1934–
  Rubberlike elasticity.

  "A Wiley-Interscience publication."
  Bibliography: p.
  Includes index.
  1. Polymers and polymerization.   2. Elasticity.
I. Erman, Burak.   II. Title.

QD381.M379 1988   547.7   88-5518
ISBN 0-471-61499-8

Printed in the United States of America

10 9 8 7 6 5 4 3 2 1

PREFACE

This book was prepared to provide a concise, elementary presentation of the most important aspects of rubberlike elasticity. Along with many of our colleagues, we have long felt a need for such an introductory treatment. The present time seems propitious because of new insights into the subject provided by theory and numerous recent developments on the experimental side.

We have treated the subject from the point of view of the physical chemist or chemical physicist. Accordingly, there is a very pronounced emphasis on molecular concepts and physical ideas, particularly those underlying some of the more abstract theory. The coverage is restricted to equilibrium properties, with no significant consideration of the huge body of literature on polymer viscoelasticity. The approach is quite tutorial, and the only background required of the reader is familiarity with the basic concepts of physical chemistry. Consequently, readers already knowledgeable about some aspects of rubberlike elasticity will be inclined to move through a few of the sections relatively rapidly. Nonetheless, we hope that all readers will benefit from the general overview and will also find some specific topics to be of particular interest and useful in their own research programs. The material presented should be sufficient for a one-term introductory course on the subject.

To a large extent, the book can be divided into two major parts. Part A deals primarily with fundamentals; Part B considers additional topics, many of which are still under intensive investigation and thus necessarily discussed in only a preliminary manner. From a positive point of view, the tentative nature of these capsule summaries should stimulate further work in these areas. For both Parts A and B, some of the more detailed material has been placed in appendixes.

Both of us had the great priviledge of collaborating extensively with the late Paul J. Flory, who contributed so remarkably to an understanding of rubberlike elasticity, among other topics in the area of polymer science. Our approach to this subject is largely his, and it would be impossible (and undesirable) to remove these partialities completely. Although they are there, every effort has been made to provide some balance by commenting on other approaches and schools of thought.

It is a great pleasure to acknowledge the invaluable assistance provided by Mrs. Jane Hershner, who typed all the drafts of the manuscript. She either has the patience of a saint or is a superb actress (probably the latter).

Finally, we wish to dedicate this book to the memory of Paul Flory, because of his seminal contributions to the area of rubberlike elasticity, but more generally on behalf of the countless people he inspired, both as a scientist and an extraordinary, profound human being.

JAMES E. MARK
BURAK ERMAN

Cincinnati, Ohio
Istanbul, Turkey
June 1988

CONTENTS

RUBBERLIKE ELASTICITY
A MOLECULAR PRIMER

PART A

FUNDAMENTALS

1

INTRODUCTION

GENERAL COMMENTS

The materials to be discussed in this book are known by a variety of names. The oldest, *rubbers*, is not very illuminating since it refers to their relatively unimportant capability of removing pencil or ink marks from paper by an abrasive rubbing action (Treloar 1975; Eichinger 1983). Of much greater importance are their elastic properties, and the term *elastomers* is now much in use. So also is *rubberlike materials*, which emphasizes the similarities between such substances and natural rubber, which is obtained from the *Hevea* tree.

Rubberlike materials have long been of extraordinary interest and importance. They find usage in items ranging from automobile tires and conveyor belts to heart valves and gaskets in supersonic jet planes. The striking nature of their elastic properties and their relationships to molecular structure has attracted the attention of numerous physical chemists and chemical physicists interested in structure–property relationships, particularly those involving polymeric materials (Flory 1953; Treloar 1975).

RUBBERLIKE ELASTICITY AND ITS MOLECULAR REQUIREMENTS

The most useful way to begin a discussion of rubberlike elasticity is to define it and then to list the molecular characteristics required to achieve the very

3

Table 1.1   Definition and Molecular Requirements for Rubberlike Elasticity

Two-Part Definition	Molecular Requirements
1. Very high deformability	1. Material constituted of molecules that are (a) long chains (polymers) (b) highly flexible and mobile
2. Essentially complete recoverability	2. Network structure from cross-linking of molecules

unusual behavior described. This is done in Table 1.1. The definition has two parts: very high deformability and essentially complete recoverability. In order for a material to exhibit this type of elasticity, three molecular requirements must be met: (1) the material must consist of polymeric chains, (2) the chains must have a high degree of flexibility and mobility, and (3) the chains must be joined into a network structure (Mark 1981, 1984a).

The first requirement is associated with the very high deformability. It arises from the fact that the molecules in an elastomeric material must be able to alter their arrangements and extensions in space dramatically in response to an imposed stress, and only a long-chain molecule has the required very large number of spatial arrangements of very different extensions. This versatility is illustrated in Figure 1.1, which depicts a two-dimensional projection of a random spatial arrangement of a relatively short polyethylene chain in the undeformed amorphous state (Mark 1981, 1984a). The spatial configuration shown was computer generated using a Monte Carlo technique in as realistic a manner as possible. The correct bond lengths and bond angles were employed, as was the known preference for trans rotational states about the skeletal bonds in any *n*-alkane molecule. A final feature taken into account was the fact that rotational states are interdependent; what one rotatable skeletal bond does depends on what the adjoining skeletal bonds are doing (Flory 1969). One important feature of this typical configuration is the relatively high spatial extension of some parts of the chain. This is due to the preference for the trans conformations, already mentioned, which are essentially planar zigzag and thus of high extension. A feature that is more important in the present context is the fact that, in spite of these preferences, many sections of the chain are quite compact. Thus the chain extension (as measured by the end-to-end separation) is quite small. For even such a short chain, the extension could be increased approximately fourfold by simple rotations about skeletal bonds, without any need for the more energy-demanding distortions of bond angles or increases in bond lengths.

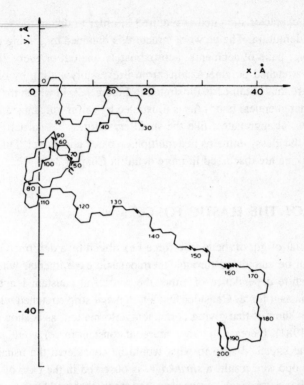

Figure 1.1 A two-dimensional projection of the backbone of an undeformed *n*-alkane chain (or sequence from a longer polyethylene chain) which consists of 200 skeletal bonds (Mark 1981, 1984a). This representative arrangement or spatial configuration was computer generated using known values of the bond lengths, bond angles, rotational angles about skeletal bonds, and preferences among the corresponding rotational states. (Reprinted with permission from J. E. Mark et al., Eds., *Physical Properties of Polymers.* Copyright 1984 American Chemical Society.)

The second characteristic required for rubberlike elasticity also relates to the high deformability. It specifies that the chains be flexible and mobile and thus that the different spatial arrangements of the chains be accessible. That is, changes in these arrangements should not be hindered by such constraints as might result from inherent rigidity of the chains, or by decreased mobility as would result from extensive chain crystallization, or from the very high viscosity characteristic of the glassy state. These two requirements are further discussed in Chapter 2, using specific examples of elastomeric and nonelastomeric materials.

The last characteristic cited is required in order to obtain the recoverability part of the definition. The network structure is obtained by joining together, or *cross-linking*, pairs of segments, approximately one out of every 100, thereby preventing stretched polymer chains from irreversibly sliding by one another. The structure thus obtained is illustrated in Figure 1.2, in which the cross-links may be either chemical bonds (as is illustrated by sulfur-vulcanized natural rubber) or physical aggregates, like the small crystallites in a partially crystalline polymer or the glassy domains in a multiphase block copolymer. Different types of cross-linking are discussed in more detail in Chapter 3.

ORIGIN OF THE ELASTIC FORCE

The molecular origin of the elastic force f exhibited by a deformed elastomeric network can be elucidated through thermoelastic experiments, which involve the temperature dependence of either the force f at constant length L or the length at constant force. Consider first a thin metal strip stretched with a weight W to a point short of that giving permanent deformation, as is shown in Figure 1.3 (Mark 1981). Increase in temperature (at constant force) would increase the length of the stretched strip in what would be considered the usual behavior. Exactly the opposite result, a *shrinkage*, is observed in the case of a stretched elastomer! For purposes of comparison, the result observed for a gas at constant pressure is included in the figure. Raising its temperature would of course cause an increase in its volume V, as is illustrated by the well-known equation $pV = nRT$.

The explanation for these observations is given in Figure 1.4 (Mark 1981). The primary effect of stretching the metal is the increase ΔE in energy caused by changing the values of the distance d of separation between the metal atoms.

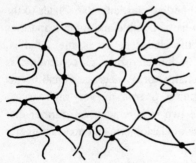

Figure 1.2   Schematic sketch of part of a typical elastomeric network.

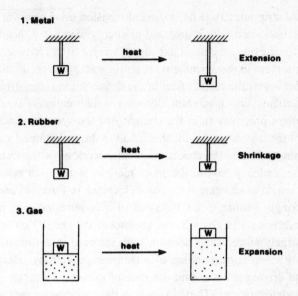

Figure 1.3 Results of thermoelastic experiments carried out on a typical metal, rubber, and gas (Mark 1981).

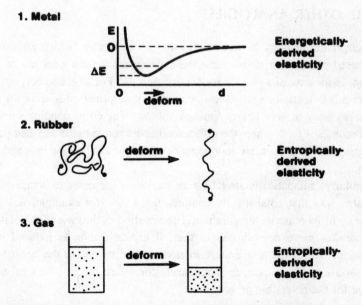

Figure 1.4 Sketches explaining the observations described in Figure 1.3 in terms of the molecular origin of the elastic force or pressure (Mark 1981).

The stretched strip retracts to its original dimension upon removal of the force since this is associated with a decrease in energy. Similarly, heating the strip at constant force causes the usual expansion arising from increased oscillations about the minimum in the asymmetric potential energy curve. In the case of the elastomer, however, the major effect of the deformation is the stretching out of the network chains, which substantially reduces their entropy. Thus, the retractive force arises primarily from the tendency of the system to increase its entropy toward the (maximum) value that it had in the undeformed state. Increase in temperature increases the chaotic molecular motions of the chains and thus increases the tendency toward the more random state. As a result there is a decrease in length at constant force, or an increase in force at constant length. This is strikingly similar to the behavior of a compressed gas, in which the extent of deformation is given by the reciprocal volume $1/V$. The pressure of the gas is largely entropically derived, with increase in deformation (i.e., increase in $1/V$) also corresponding to a decrease in entropy. Heating the gas increases the driving force toward the state of maximum entropy (infinite volume or zero deformation). Thus, increasing the temperature increases the volume at constant pressure, or increases the pressure at constant volume.

SOME OTHER ANALOGIES

The surprising analogy between a gas and an elastomer (which after all is a condensed phase) carries over into the expressions for the work dw of deformation. In the case of a gas, dw is of course $-p\,dV$. For an elastomer, however, this term is essentially negligible since network deformation (except for swelling) takes place at very nearly constant volume. The corresponding work term now becomes $+f\,dL$, where the difference in sign is due to the fact that positive w corresponds to a decrease in volume of a gas but an increase in length of an elastomer.

Similarly, adiabatically stretching an elastomer increases its temperature in the same way that adiabatically compressing a gas (for example, in a diesel engine) will increase its temperature. The situation in the case of the elastomer is somewhat more complicated in that, if crystallization is induced by the stretching, part of the temperature increase would be due to the latent heat of crystallization. In any case, for both the elastomer and the gas, the total entropy change for the reversible process

$$\Delta S = \Delta S(\text{deformation}) + \Delta S(\text{temperature change}) \qquad (1.1)$$

must be positive or zero since the systems are acting as though they are (temporarily) isolated. Thus, since ΔS(deformation) must be negative, ΔS(temperature change) must be positive, and this has to correspond to a temperature increase. The basic point here is the fact that the retractive force of an elastomer and the pressure of a gas are both primarily entropically derived, and as a result the thermodynamic and molecular descriptions of these otherwise dissimilar systems are very closely related.

As would be expected, letting a stretched elastomer contract adiabatically causes its temperature to decrease. An analogy of this effect is provided by adiabatic demagnetization, a technique used to reach very low temperatures (Atkins 1982). A suitable salt is magnetized isothermally by the application of a strong field, thereby aligning its magnetic moments with an associated decrease in entropy. The field is then removed adiabatically (which is analogous to letting the elastomer snap back), and the moments again spontaneously become disordered. In both cases ΔS(disordering) is positive and is offset by a negative ΔS(temperature change), which of course corresponds to a decrease in temperature.

The fact that heat is given off in the stretching of an elastomer can be used to provide a thermodynamic explanation of the observed shrinkage of a stretched elastomer when its temperature is increased. According to Le Châtelier's principle, "A system at equilibrium, when subjected to a perturbation, responds in a way that tends to eliminate the effect" (Atkins 1982). Thus, since heat is given off during stretching, adding heat has to cause a contraction.

The temperature increase observed upon stretching an elastomer is augmented by heat generated by the wasteful conversion of part of the deformation energy through frictional effects. Thus the temperature increase during the stretching process is not completely offset by the temperature decrease during the retraction phase. There is therefore a hysteretic buildup in temperature that is highly disadvantageous. Not only does it represent wastage of mechanical energy, but the heat buildup can have a degradative effect on the elastomer. Probably the most important example here is the flexing of an automobile tire as it rotates through its bending–recovery cycles.

SOME HISTORICAL HIGH POINTS

Table 1.2 summarizes some important contributions from early experiments on rubberlike elasticity. The earliest experiments, by Gough in 1805 (Flory 1953; Treloar 1975; Mason 1979), demonstrated the heat effects described in the pre-

Table 1.2   Some Early Contributions in the Experimental Area

Contribution	Scientist	Date
Heat effects, strain-induced crystallization	Gough	1805
Vulcanization (cross-linking)	Goodyear and Hayward	1839
Thermoelasticity	Joule	1859
Volume changes accompanying deformation	Several	~1930

ceding few paragraphs and also the phenomenon of strain-induced crystallization, which is discussed in Chapter 12. The discovery of *vulcanization* (cross-linking) by Goodyear and Hayward in 1839 greatly facilitated experimental investigations since cross-linked elastomers could now be brought close to elastic equilibrium (Morawetz 1985). The more quantitative thermoelasticity experiments described above were first carried out by Joule back in 1859. This was in fact only a few years after entropy was introduced as a thermodynamic function. Another important experimental contribution was the observation by several workers that deformations (other than swelling) of rubberlike materials occurred essentially at constant volume, so long as crystallization was not induced (Gee et al. 1950). (In this sense, the deformation of an elastomer and a gas are very different.)

Some early contributions on the theoretical side are described in Table 1.3. A molecular interpretation of the fact that rubberlike elasticity is primarily entropic in origin had to await Hermann Staudinger's much more recent demonstration that polymers were covalently bonded molecules and not some type of association complex best studied by the colloid chemists (Morawetz 1985). Meyer, von Susich, and Valko in 1932 correctly interpreted the observed near constancy in volume to indicate that the changes in entropy must therefore involve changes in orientation or configuration of the network chains. They also concluded that the elastic force should be proportional to the absolute temperature (Treloar 1975). These basic qualitative ideas are shown in the sketch in Figure 1.5, where the arrows represent some typical end-to-end vectors of the network chains. The first step toward making these ideas more quantitative, in the form of an elastic equation of state, was the idea, proposed by Werner Kuhn in 1936, that the elastic force f should be proportional to the number of "molecules" in the elastomer. In the 1930s, Kuhn, Eugene Guth, and Herman Mark first began to develop quantitative theories based on this idea that the network chains undergo configurational changes, by skeletal bond rotations, in response

Table 1.3   Some Early Contributions in the Theoretical Area

Contribution	Scientist	Date
Chainlike nature of polymers	Staudinger	~1920
Chain orientation upon network deformation; elastic force proportional to absolute temperature	Meyer, von Susich, and Valko	1932
Elastic force proportional to number of "molecules"	Kuhn	1936
Elastic force proportional to absolute temperature and to sample length	Guth and Mark	1934
Phantom Network theory	James and Guth	1941
Affine Network theory	Wall	1942
Affine Network theory	Flory and Rehner	1943

to an imposed stress. Guth and Mark (1934) also concluded from this picture that f should be proportional to absolute temperature, which turns out to be approximately correct. They also concluded, however, that f should be proportional to the length of the stretched elastomer. This is incorrect since the constant-volume nature of the elongation process requires that the sample dimensions perpendicular to the stretching direction *decrease* proportionally. Some chains are therefore compressed in an elongation experiment, as is illustrated

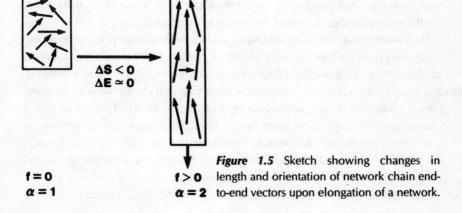

$\Delta S < 0$
$\Delta E \approx 0$

$f = 0$
$\alpha = 1$

$f > 0$
$\alpha = 2$

Figure 1.5 Sketch showing changes in length and orientation of network chain end-to-end vectors upon elongation of a network.

by the horizontal end-to-end vector shown in the middle of the sample strip in Figure 1.5. As a result, f is not proportional to L or to the elongation $\alpha = L/L_i$ (L_i being the initial length), but to $(\alpha - \alpha^{-2})$, where the subtractive term $-\alpha^{-2}$ results from these compressive effects.

Guth, in collaboration with Hubert James, began development of the *phantom network theory* of rubberlike elasticity around 1941. Frederick Wall, Paul Flory, and John Rehner in 1942 and 1943 then began development of the alternative *affine network theory*. Both theories are described in Chapter 4.

BASIC POSTULATES

There are two postulates that have been of critical importance in the development of molecular theories of rubberlike elasticity. The most important is:

1. Although *inter*molecular interactions unquestionably occur in rubberlike materials (they are, after all, condensed phases), these interactions are independent of configuration. They are thus independent of extent of deformation and therefore play no role in deformations carried out at constant volume and composition, except of course in the case of strain-induced crystallization.

The postulate states that rubberlike elasticity is an *intra*molecular effect, specifically the entropy-reducing orientation of network chains described above. As a repercussion, these chains should be random in the bulk (undiluted) amorphous state, in the absence of any deformation (Flory 1953). Since intermolecular effects are independent of intramolecular effects, there is no inducement for the spatial configurations of the chains to be altered.

This assumption, probably originally made more out of desperation than anything else, is now supported by a variety of results. First, thermoelastic results are found to be independent of network swelling, as is described in Chapter 9. Second, neutron scattering studies have confirmed that chains in the bulk, amorphous, undeformed state are indeed random (Flory 1984). They in fact have mean-square dimensions the same as the unperturbed values $\langle r^2 \rangle_0$ they have in a θ-system, where excluded-volume effects are nullified (Flory 1953). Finally, the partition function for the system has been found to be separable (to factor) into an intramolecular part and a compositional part (Flory 1984).

The second postulate is very closely related to the first. It states:

2. The Helmholtz free energy of the network should also be separable:

$$A = A_{\text{liq}}(T, V, N) + A_{\text{el}}(\lambda) \qquad (1.2)$$

where λ is the strain tensor.

It is thus assumed that the nonelastic (liquidlike) part of the network free energy is independent of deformation. How much of an approximation this involves is not yet entirely clear. Differential sorption measurements on networks may eventually resolve this issue (Yen and Eichinger 1978).

In some of the theories, it is further assumed that the deformation is affine, that is, that the network chains move in proportion to the macroscopic dimensions of the elastomeric sample (Flory 1953; Treloar 1975; Eichinger 1983). This assumption was relaxed in the phantom network theories and in the more modern theory described in Chapter 5.

It is also frequently assumed that the chains have end-to-end distances that follow a Gaussian distribution. This is quite satisfactory except of course for chains that are unusually stiff, very short, or brought close to the limits of their extensibility by the deformation process. Non-Gaussian behavior is described in Chapter 13.

A TYPICAL STRESS–STRAIN APPARATUS

The apparatus typically used to measure the force required to give a specified elongation in a rubberlike material is very simple, as can be seen from its schematic description in Figure 1.6 (Mark 1981, 1984a). The elastomeric strip is mounted between two clamps, the lower one fixed and the upper one attached to a movable force gauge. A recorder is used to monitor the output of the gauge as a function of time in order to determine when the force is exhibiting the constant, near-equilibrium values suitable for comparisons with theory. The sample is generally protected with an inert atmosphere such as nitrogen to prevent degradation, particularly in the case of measurements carried out at elevated temperatures. Both the sample cell and container with the surrounding constant-temperature bath are made of glass, thus permitting use of a cathetometer or traveling microscope to obtain values of the strain by measuring the distance between two lines marked on the central portion of the test sample. Thermoelastic measurements are of course carried out by changing the temperature of the constant temperature bath.

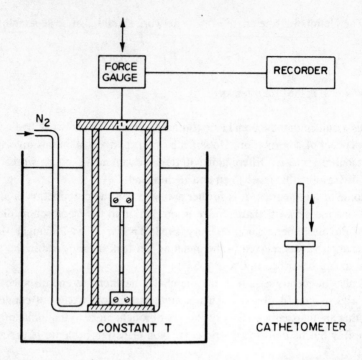

Figure 1.6 Apparatus for carrying out stress–strain measurements on an elastomer in elongation (Mark 1981, 1984a). (Reprinted with permission from J. E. Mark et al., Eds., *Physical Properties of Polymers*. Copyright 1984 American Chemical Society.)

A TYPICAL STRESS–STRAIN ISOTHERM

A typical stress–strain isotherm obtained on a strip of cross-linked natural rubber as described above is shown in Figure 1.7 (Mark 1981, 1984a). The units for the force are generally newtons (N), and for the nominal or engineering stress f^* (the force divided by the undeformed cross-sectional area) they are N mm^{-2}. The curves obtained are usually checked for reversibility. In this type of representation, the area under the curve is frequently of considerable interest since it is proportional to the work of deformation $w = \int f\,dL$. Its value up to the rupture point is thus a measure of the toughness of the material.

The initial part of the stress–strain isotherm shown is of the expected form in that f^* approaches linearity with elongation α as α becomes sufficiently large to make negligibly small the α^{-2} term in the $(\alpha - \alpha^{-2})$ strain function mentioned above.

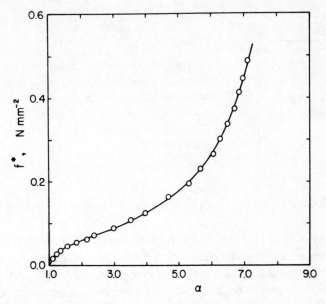

Figure 1.7 Stress-elongation curve for natural rubber in the vicinity of room temperature (Mark 1981, 1984a). (Reprinted with permission from J. E. Mark et al., Eds., *Physical Properties of Polymers*. Copyright 1984 American Chemical Society.)

The large increase in f^* at high deformation in the case of natural rubber is due largely if not entirely to strain-induced crystallization. The melting point $T_m = \Delta H_m / \Delta S_m$ of the polymer is directly proportional to the heat of fusion ΔH_m and inversely proportional to the entropy of fusion ΔS_m. The latter is significantly diminished (from ΔS to $\Delta S'$) when the chains in the amorphous (melted) network remain stretched out because of the applied deformation. The melting point is thereby increased, and it is in this sense that the stretching induces the crystallization of some of the network chains. Figure 1.8 (Mark 1984b) illustrates these ideas. The effect is a mechanical analog to suppressing the extent to which chains can randomize themselves by putting rigid groups such as p-phenylene in the polymeric repeat unit, as is described in Chapter 12 (Mark 1984b). It is also qualitatively similar to the increase in melting point generally observed upon increase in pressure on a low-molecular-weight substance in the crystalline state (Atkins 1982). In any case, the crystallites thus formed act as physical cross-links and reinforcing filler, and they decrease the amount of deformable material in the sample (Mark 1979a,b). These changes increase the modulus of the network. The properties of both crystallizable and

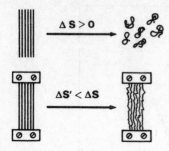

Figure 1.8 Sketch explaining the increase in melting point with elongation in the case of a crystallizable elastomer. (Reprinted with permission from J. E. Mark and G. Odian, Eds., *Polymer Chemistry*. Copyright 1984 American Chemical Society.)

noncrystallizable networks at high elongations are discussed in detail in Chapter 12.

Additional deviations from theory are found in the region of moderate deformation upon careful examination of the data, particularly in terms of the usual plots of modulus against reciprocal elongation. Although early theories predicted the modulus to be independent of elongation, it generally decreases significantly upon increase in α. The interpretation of these deviations is given in Chapter 8.

SCOPE OF THE COVERAGE OF THE SUBJECT

As already mentioned in the Preface, the focus will be almost entirely on rubberlike materials at elastic equilibrium. Nonequilibrium, viscoelastic properties have been extensively discussed in a number of other places (Ferry 1980; Aklonis and MacKnight 1983). Interpretations will be in molecular terms, and nonmolecular (phenomenological) discussions will have to be sought elsewhere (Treloar 1975; Gee 1980; Ogden 1986). Homogeneous (one-phase) systems will be of primary interest, the major exceptions being elastomers undergoing strain-induced crystallization and filled elastomers for which there is some molecular understanding.

2

SOME RUBBERLIKE MATERIALS

As mentioned in Chapter 1, the preparation of an elastomeric material requires that the polymer chains being cross-linked be relatively long and have significant flexibility and mobility. Some examples (Morton 1973; Treloar 1975; Elias 1977) of polymers meeting these requirements are given in Table 2.1. As can be seen from the structures given, these typical chains have few double bonds and no rings as part of the backbone repeat unit. This is to be expected, since such structural features decrease the flexibility of a polymer. The high flexibility of these chains is shown in part by the relatively low values they exhibit for the *glass transition temperature* T_g, which is the temperature below which flexibility is so reduced that the material becomes glassy.

Natural rubber, which is essentially 100% of the *cis*-1,4 form of polyisoprene, has a relatively low value of T_g. Specifically, noncrystalline regions of this polymer will not lose their rubberlike elasticity until the temperature falls below $-73°C$. Significant amounts of crystallinity in the undeformed state interfere with elastomeric behavior since the chain segments packed into crystallites obviously have their flexibility suppressed. The maximum equilibrium melting point of natural rubber is $28°C$, but T_m is typically depressed significantly when the polymer is compounded with various additives in a commercial elastomer. Therefore, as a practical matter, natural rubber is a good elastomer and is much used, despite the fact that room temperature is below the maximum melting point of the polymer.

Table 2.1   Some Polymers That Are Normally Rubberlike

Polymer	Structure	T_g (°C)	T_m (°C)
Natural rubber[a]	$[C(CH_3)=CH-CH_2-CH_2-]$	−73	28
Butyl rubber[b]	$[C(CH_3)_2-CH_2-]$	−73	5
Poly(dimethylsiloxane)	$[Si(CH_3)_2-O-]$	−127	−40
Poly(ethyl acrylate)[c]	$[CH(COOC_2H_5)-CH_2-]$	−24	—
Styrene–butadiene co-polymer	$[CH(C_6H_5)-CH_2-]$, $[CH=CH-CH_2-CH_2-]$	Low	—
Ethylene–propylene copolymer	$[CH_2-CH_2-]$, $[CH(CH_3)-CH_2-]$	Low	—

[a]*Cis*-1,4-polyisoprene.
[b]Polyisobutylene containing a few mole percent unsaturated comonomer.
[c]Stereochemically irregular (atactic) polymer.

Another important elastomer is butyl rubber, which coincidentally has the same low value of T_g as natural rubber. Its melting point of 5°C is, however, well below room temperature. It has one of the lowest permeabilities of any elastomer and is therefore used extensively in pneumatic applications, such as the inner lining of tires.

Poly(dimethylsiloxane) (PDMS), a semi-inorganic polymer, is used in many specialty applications, particularly those requiring a high-performance elastomer. It remains noncrystalline and elastomeric to very low temperatures ($T_m \simeq -40$°C), and its excellent thermal stability permits usage at unusually high temperatures. Its glass transition temperature, −127°C, is the lowest reported for any polymer. The very high chain flexibility this indicates is due to (1) the relatively long Si–O bond (1.64 Å), (2) the unusually large bond angle (143°) at the O atoms, and (3) the small size of the totally unencumbered O atoms. It is only the methyl groups that distinguish this material from the totally inorganic silicates such as window glass. Therefore, not surprisingly, the polymer is highly inert and can be used in such demanding applications as high-temperature seals and body implants.

Since a large amount of crystallinity in the undeformed state is disadvantageous, some polymers are prepared so as to make them *inherently* noncrystallizable. In the case of poly(ethyl acrylate), this is done by using a polymerization initiator or catalyst that does not control the stereochemical structure at the substituted (pseudo-asymmetric) carbon atoms in the chain backbone. The resulting atactic polymer is incapable of crystallizing because of its stereochemical irregularities. A polymer chain can also be made structurally irregular by preparing it in a random copolymerization. Styrene–butadiene and ethylene–propylene copolymers are in this category, and neither typically has a significant amount of crystallinity. Their specific values of T_g depend of course on chemical composition, but they are known to be low. Thus the two chemical copolymers and the stereochemical copolymer are rubbery and are in fact commercially important elastomers.

Although large amounts of crystallinity in the undeformed state is undesirable, it should be mentioned that the generation of some crystallites *during* the stretching process can greatly increase the toughness of an elastomer. This type of *strain-induced crystallization* is discussed in Chapter 12.

Some typical polymers that are not usually rubberlike are described in Table 2.2. In the case of polyethylene, the chains are inherently flexible, but a high degree of crystallinity in the undiluted (unswollen) state at room temperature reduces their mobility and thus prevents the occurrence of rubberlike elasticity. This crystallinity can be removed by either increasing the temperature to above the melting point of the polymer ($\sim 130°C$), or by incorporating a swelling diluent (plasticizer). Elastomeric measurements have in fact been carried out on rubbery cross-linked polyethylene, both at high temperatures in the unswollen state and at somewhat lower temperatures when swollen with a nonvolatile diluent. These experiments on network thermoelasticity are described in Chapter 9. They are not as peculiar as it might seem since it was assumed, correctly, that the unusually simple structure of this polymer would facilitate theoretical interpretation of the experimental results.

Similarly, cross-linked polystyrene (which is normally glassy) can be made elastomeric either by increasing the temperature to above its value of T_g ($\sim 100°C$) or by incorporating a plasticizer. Again, such experiments have ac-

Table 2.2   Some Polymers That Are Not Normally Rubberlike

Polymer	Structure	Complication
Polyethylene	$[CH_2-CH_2-]$	Highly crystalline
Polystyrene	$[CH(C_6H_5)-CH_2-]$	Glassy
Poly(vinyl chloride)	$[CHCl-CH_2-]$	Glassy
Elastin	$[CO=NH-CHR-]$	Glassy
Polymeric sulfur	$[S-]$	Chains too unstable
Poly(p-phenylene)	$[C_6H_4-]$	Chains too rigid
Bakelite®[a]	$[C_6H_4(OH)-CH_2-]$, etc.	Chains too short

[a]Phenol–formaldehyde resin.

tually been done, but in this case primarily because a convenient polymerization reaction of styrene was known to give model networks of known structure (Herz et al. 1978). Poly(vinyl chloride) provides a more practical example. Its rather high value of T_g ($\sim 85°C$) is frequently suppressed to a value below room temperature by means of a nonvolatile plasticizer, making it sufficiently elastomeric for a number of applications.

An example from the realm of biopolymers is provided by the protein elastin, whose repeat units have different R side groups, as shown in Table 2.2 This bioelastomer is used by mammals, including man, for a variety of elastomeric applications (Gosline 1987). The dry polymer has a glass transition temperature in the vicinity of 200°C (according to extrapolations of data), which is much too high for living systems. Nature, however, also apparently knows about plasticizers; it never uses elastin except as sufficiently highly swollen by aqueous body fluids to bring its T_g value well below the operating temperature of mammals, $\sim 37°C$ (98.6°F). This fascinating material is discussed further in Chapter 19.

If elemental sulfur is melted, polymerization can occur at these elevated temperatures, and the resulting chains do exhibit some rubbery behavior in the quenched state. The material has never been made truly rubber-elastic, however. The chains could not be cross-linked and, because of their instability, could not be kept from reorganizing. Therefore, the required recoverability could not be achieved.

There are two general classes of polymers that are inherently incapable of exhibiting rubberlike elasticity. Rigid-rod polymers such as poly(p-phenylene) lack the required flexibility because of their chemical structure. Also, very highly cross-linked thermosets, such as the phenol–formaldehyde resins (Bakelite®) or the epoxy adhesives, generally consist of chains that are much too short to have the required flexibility under any conditions.

3

PREPARATION AND STRUCTURE OF NETWORKS

PREPARATION OF NETWORKS

A network is obtained by permanently linking polymeric chains together in the form of a three-dimensional structure such as is shown in Figure 1.2. The process is known as cross-linking, curing, or vulcanization (Stephens 1973; Coran 1978). The points of linking, called *junctions*, may be (1) randomly located along the chains, or (2) found at specific locations such as the ends or selected repeat units [particularly in the case of bioelastomers (Gosline 1987)]. The required linking or *cross-linking* can be brought about by either chemical reactions that form covalent bonds between chains, or by physical aggregation of units from two or more chains. The simplicity of the first technique and the permanence of the structures it provides are advantages for some applications. The main advantage of the second technique is the reprocessability of the elastomer, which results from the fact that the aggregation process is generally reversible.

The most important examples of cross-linking using chemical reactions that attack the chains at random locations are given in Table 3.1. Sulfur is used primarily with elastomers having numerous unsaturated groups, for example, those in the repeat units of the polyisoprenes and polybutadienes. Sulfur adds to some of these double bonds (approximately 1%), resulting in a short chain of x sulfur atoms joining two chains of elastomer as shown in the sketch (Coran 1978).

Table 3.1   Cross-Linking by Random Chemical Reactions

A. *Sulfur cures*

$$2 \sim\sim + xS \longrightarrow \; \underset{|}{\overset{|}{S_x}}$$

B. *Peroxide cures*

Peroxide $\longrightarrow$ 2R$\cdot$

R$\cdot$ + $\sim$CH$_2$$\sim$ $\longrightarrow$ RH + $\sim$ĊH$\sim$

$$2 \sim\text{ĊH}\sim \longrightarrow \; \overset{\sim\text{CH}\sim}{\underset{\sim\text{CH}\sim}{|}}$$

C. High-energy radiation (e, γ, UV)

$$2 \sim\text{CH}_2\sim \longrightarrow \text{H}_2 + \overset{\sim\text{CH}\sim}{\underset{\sim\text{CH}\sim}{|}}$$

Peroxide cures can be used for a greater variety of elastomers. A suitably unstable peroxide is homogeneously cleaved into two free radicals R$\cdot$, each of which then abstracts a hydrogen radical from the polymer chain to become the stable molecule RH. The free radical now present on the polymer migrates along the chain until it is in proximity to a similar radical on another chain. Combination of the two radicals results in a covalent bond that cross-links the two chains (Flory 1953). The radicals formed from some peroxides are of relatively low reactivity and in order to function must find groups such as vinyl side chains that are more reactive than methylene or methyl groups.

The third random technique involves high-energy radiation, such as electrons (e), gamma photons (γ), and ultraviolet (UV) light (Dole 1972). Free radicals are formed and function as described above, but also of importance is the generation of ions (hence the synonym *ionizing radiation*). This is the most abusive of the techniques and generally also causes a great deal of chain scission, resulting in dangling-chain irregularities in the network structure.

Network structures may also be formed by the random copolymerization of monomers, at least one type of which has a *functionality* ϕ of 3 or greater, where ϕ is the number of sites from which chains can grow. An example is the condensation of dibasic acids ($\phi = 2$) with glycerol ($\phi = 3$) (Flory 1953). The reaction typically takes place in solution, with the system going from a liquid to a solid, specifically to a network structure swollen with any unreacted monomers and whatever solvent was present. The process is called *gelation*, and the resulting swollen network is a *gel*. The molecular aspects of gelation in general have been extensively investigated from the theoretical point of view (Flory 1953; Leung and Eichinger 1984; Dušek 1986; Stepto 1986; Miller and Macosko 1987). This terminology has been extended in that the term *gel* is now frequently applied even to polymer networks that are swollen by absorption of a diluent after they have been cross-linked in the dry state. Swollen polymer networks are discussed further in Chapters 7, 8, and 17.

There are some polymers that cannot be cured by these relatively simple techniques, polyisobutylene being a good example. In order to obtain a curable polymer, isobutylene monomer is copolymerized with a comonomer, which results in sites of unsaturation along the chain (Zapp and Hous 1973). These sites are then susceptible to the techniques mentioned above. The same approach is used for polymers that are difficult to cure, an example being the terpolymerization of ethylene, propylene, and diene monomers to yield a curable ethylene–propylene elastomer (Borg 1973).

Cross-linking by highly specific chemical reactions is illustrated in Table 3.2 (Gottlieb et al. 1981; Mark 1982a; Leung and Eichinger 1984; Dušek 1986; Miller and Macosko 1987). For example, a chain having hydroxyl groups at both ends can be end-linked into a network through either an addition reaction with a triisocyanate $[Y(N{=}C{=}O)_3]$, or through a condensation reaction with a trifunctional or tetrafunctional alkoxysilane [e.g., $Si(OC_2H_5)_4$]. Some of these reactions are discussed further in Chapter 10. Since the reactive groups are only at the chain ends in this example, the end-linking reactant has to have a functionality ϕ of 3 or greater. [The case where $\phi = 2$ merely gives chain extension (Mark and Sung, 1982) rather than cross-linking.]

The copolymerization approach mentioned above is an example of reacting side chains to give the desired cross-links. Since the chains themselves have a

Table 3.2   Cross-Linking by Highly Specific Chemical Reactions

A. End-group reactions

   X⌁X + reactant having $\phi \geq 3$

B. Side-chain reactions

$$
\begin{array}{c}
X \\
| \\
{\sim}{\sim}{\sim}{\sim}{\sim}{\sim}{\sim}{\sim}{\sim}{\sim} \quad + \text{ reactant having } \phi \geq 2 \\
| \qquad\quad | \\
X \qquad\quad X
\end{array}
$$

very high functionality, the reactant can have a functionality as low as 2 and still cure the polymer into a network structure. Another example of this approach is given in Chapter 19.

Some examples of cross-linking by the physical aggregation of chains are given in Table 3.3. In case A, the elastomer chains are absorbed onto the surface of a finely divided particulate filler of the type used to reinforce elastomers (Boonstra 1979). Examples would be carbon black added to natural and various synthetic rubbers, and silica added to silicone elastomers. These and other fillers are discussed in Chapter 20. Of interest here is the fact that reactive groups on the particle surfaces can easily make the adsorption strong enough to effectively cross-link the chains, generally in an essentially irreversible manner (Warrick et al. 1979). It can occur so fast that the material sets (gels) prematurely, thus interfering with its processing. In such cases, some of the reactive groups on the particle surfaces are deactivated by treatment with a low-molecular-weight additive prior to the blending of the filler into the polymer.

The second example involves polymers having a relatively small number of very small crystallites, for example plasticized poly(vinyl chloride) (Davis

Table 3.3   Cross-Linking by Physical Aggregation

A. Filled polymers	B. Microcrystalline polymers

C. Ionomers	D. Chelation polymers[a]
M^+	acac acac Pd^{2+} Pd^{2+} acac acac

E. Triblock copolymers[b]

S—∿—S—BD—∿—BD—S—∿—S

S—∿—S—BD—∿

[a] acac = acetylacetonate.
[b] BD = butadiene.

1973). As shown in the sketch (Table 3.3B), these microcrystallites act as cross-links of very high functionality. Since this type of aggregation is obviously reversible by heating above the melting point of the polymer, these elastomeric materials are reprocessable.

Ionomers are polymers containing a small number of ionic side chains (typically a few mole percent) (Eisenberg and King 1977). These ions can be made to aggregate around a metal ion carrying one or more positive charges (M^+), as shown in the sketch (Table 3.3C). This type of cross-linking (Bagrodia et al. 1987) is also reversible, by increasing the temperature or by adding a solvent. In a very similar technique (Table 3.3D), acetylacetonate side groups on a polymer have been chelated to a metal atom such as palladium (Yeh et al. 1982). Reversal of this curing step is brought about by adding other chelating substances to compete with the groups attached to the polymer chains.

The final example pertains specifically to triblock copolymers (Holden 1973; Noshay and McGrath 1977) where the first and third blocks can form hard glassy or crystalline domains. In the sketch shown in Table 3.1E (Mark 1984b), these two blocks consist of the glassy polymer polystyrene and are separated by a block of the elastomeric polymer polybutadiene. The two types of blocks do not mix because of the very low entropy of mixing chainlike molecules. Phase separation thus occurs and is held at the microscopic level by the covalent bonding going through the polystyrene–polybutadiene interfaces. As a result the styrene sequences segregate into domains having diameters the order of 200 Å, as shown in the table and in Figure 3.1 (Mark 1984b), where the left portion of the sketch shows some of the molecular detail. The domains act as crosslinks, but the process is obviously reversible. In the example shown, the material would be reprocessable at a temperature above the glass transition temperature of polystyrene ($\sim 100°C$). In the case where the hard domains are crystalline, this temperature would have to be above their melting point. Re-

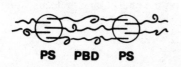

PS   PBD   PS

Figure 3.1 Phase separation giving hard polystyrene domains that effectively cross-link a polybutadiene matrix into a thermoplastic elastomer. PS, polystyrene; PBD, polybutadiene. (Reprinted with permission from J. E. Mark and G. Odian, Eds., *Polymer Chemistry.* Copyright 1984 American Chemical Society.)

processable materials of this type are called *thermoplastic elastomers*, meaning they become "plastic" (moldable) upon increase in temperature.

STRUCTURE OF NETWORKS

A *network chain* (a chain between two junction points) forms the basis of the elementary molecular theory of amorphous polymeric networks (Flory 1953). In general, network chains exhibit a distribution of molecular weights about an average, which serves as a representative reference quantity in describing network structure. As already mentioned, the number of chains meeting at each junction is called the *functionality* ϕ of that junction. A network may have two or more sets of junctions with different functionalities; it may then be characterized by an average functionality. A chain connected to a junction of the network at only one end is called a *dangling chain*, and one having both of its ends attached to the same junction is called a *loop*. A network with no dangling chains or loops and in which all junctions have a functionality greater than 2 is called a *perfect network*. Although a perfect network can never be obtained in reality, it forms a simple reference structure around which molecular theories are constructed. For this reason, most of the discussion in this text is based on perfect networks.

The topological structure of a perfect network may be described by various parameters: the average molecular weight between junctions, M_c; the average functionality ϕ; the number of network chains ν; the number of junctions μ; and the *cycle rank* ξ, which denotes the number of chains that have to be cut in order to reduce the network to a tree with no closed cycles. These five parameters are not independent, however, and are related by three equations.

The first relation, between ν and μ, is very simple to derive (Flory 1953; Mark 1982b). Since in the case of a perfect network the method used to form it is irrelevant, it is easiest to consider the process as an end-linking. For the network to be perfect, the number of chain ends, 2ν, has to be equal to the number of functional groups, $\phi\mu$, on the end-linking molecules. Thus,

$$\mu = 2\nu/\phi \tag{3.1}$$

which states, for example, that to get ν chains in a perfect tetrafunctional network, only $\mu = \nu/2$ junctions are required. This can easily be seen pictorially from the sketches in Figure 3.2, where the networks are not only perfect but also simple in the sense of having a small-enough number of chains and cross-

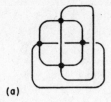
(a)

(b)

Figure 3.2 Sketches of some simple, perfect networks having (a) tetrafunctional and (b) trifunctional cross-links (both of which are indicated by the heavy dots) (Mark 1982b).

links to be easily counted. In the tetrafunctional network in Figure 3.2*a*, the eight network chains require four cross-links, as expected. On the other hand, the six chains in the trifunctional network shown in Figure 3.2*b* require the same number of junctions, since the conversion factor $2/\phi$ is now two-thirds instead of one-half.

The other two equations are

$$\xi = (1 - 2/\phi)\nu \tag{3.2}$$

$$\xi/V_0 = \frac{(1 - 2/\phi)\rho}{M_c/N_A} \tag{3.3}$$

where V_0 is the volume of network in the state of formation, ρ is the corresponding density, and N_A is Avogadro's number. Equations (3.2) and (3.3) are derived in Appendix A, along with a discussion of the modifications they require for the case of imperfect networks.

4

ELEMENTARY STATISTICAL THEORY FOR IDEALIZED NETWORKS

As was mentioned in Chapter 1, the elementary molecular theory of polymer networks rests on the postulate that the elastic free energy of a network is equal to the sum of the elastic free energies of the individual chains. Therefore, intermolecular contributions to the total elastic free energy are assumed to be insignificant and are entirely neglected (Flory 1953; Treloar 1975). Understanding of the theory thus requires a precise description of the statistical behavior of the individual chains and of the relationship between their dimensions and the macroscopic strain.

The chains in a network formed in the amorphous bulk state exhibit unperturbed dimensions identical to that of a single chain in θ-solvents (Flory 1953). This results from the fact that the distribution of the end-to-end vectors $\mathbf{r}$ for the chains in the bulk state is unchanged upon formation of network junctions, i.e., extended chains in the bulk state are equally susceptible to the interlinking or cross-linking reaction as others. The distribution of the end-to-end vector $\mathbf{r}$ of the chains in the network may therefore be identified with that of the single free chain. For sufficiently long chains, typically with 100 or more skeletal bonds, the distribution may satisfactorily be approximated by a Gaussian function:

$$W(\mathbf{r}) = (3/2\pi \langle r^2 \rangle_0)^{3/2} \exp(-3r^2/2\langle r^2 \rangle_0) \tag{4.1}$$

Here, $\langle r^2 \rangle_0$ denotes the mean-squared value of $\mathbf{r}$ for an unperturbed free chain.

The thermodynamic expression relating the elastic free energy A_{el} of a chain having an end-to-end vector $\mathbf{r}$, to $W(\mathbf{r})$ is

$$A_{el} = c(T) - kT \ln W(\mathbf{r}) \tag{4.2}$$

where $c(T)$ is a function of absolute temperature T, and k is the Boltzmann constant. Substituting Eq. (4.1) into Eq. (4.2) leads to the elastic free energy of a chain at a fixed $\mathbf{r}$ in the network

$$A_{el} = A^*(T) + (3kT/2\langle r^2 \rangle_0)r^2 \tag{4.3}$$

where $A^*(T)$ is a function only of T.

The total elastic free energy ΔA_{el} of the network relative to the undeformed state is obtained by summing Eq. (4.3) over the ν chains of the network:

$$\begin{aligned}
\Delta A_{el} &= \frac{3kT}{2\langle r^2 \rangle_0} \sum_{\nu} (r^2 - \langle r^2 \rangle_0) \\
&= \frac{3}{2} \nu kT \left(\frac{\langle r^2 \rangle}{\langle r^2 \rangle_0} - 1 \right)
\end{aligned} \tag{4.4}$$

The quantity $\langle r^2 \rangle = \sum_{\nu} r^2/\nu$ represents the average of the mean-squared end-to-end chain vectors in the deformed network.

The relationship between the mean-squared chain dimensions $\langle r^2 \rangle$ in the deformed network and $\langle r^2 \rangle_0$ is required for further development of the theory. A molecular model has to be adopted for such a relationship. The affine and the phantom network models are the two simplest molecular models employed in relating the deformation of the chains to macroscopic deformation.

The macroscopic state of deformation may be assumed to be homogeneous, where the principal extension ratios in the laboratory-fixed coordinate system $0xyz$ are defined by

$$\lambda_x = L_x/L_{x0}, \quad \lambda_y = L_y/L_{y0}, \quad \lambda_z = L_z/L_{z0} \tag{4.5}$$

Here L_x, L_y, L_z, and L_{x0}, L_{y0}, L_{z0} denote the deformed and undeformed macroscopic dimensions of a prismatic test sample.

In both the affine and the phantom network models, the chains are assumed to be of equal length. Thus polydispersity is not taken into account. Average chain dimensions are represented in the undeformed state by

$$\langle r^2 \rangle_0 = \langle x^2 \rangle_0 + \langle y^2 \rangle_0 + \langle z^2 \rangle_0 \tag{4.6}$$

and in the deformed state by

$$\langle r^2 \rangle = \langle x^2 \rangle + \langle y^2 \rangle + \langle z^2 \rangle \tag{4.7}$$

where the angular brackets indicate averaging over all chains of the network at a given instant. Assuming an isotropic network in the state of rest further leads to

$$\langle x^2 \rangle_0 = \langle y^2 \rangle_0 = \langle z^2 \rangle_0 = \langle r^2 \rangle_0 / 3 \tag{4.8}$$

THE AFFINE NETWORK MODEL

The junction points in the *affine network model* (Flory 1953) are assumed to be embedded in the network. As a result of this assumption, components of each chain vector transform linearly with macroscopic deformation:

$$x = \lambda_x x_0, \quad y = \lambda_y y_0, \quad z = \lambda_z z_0 \tag{4.9}$$

$$\langle x^2 \rangle = \lambda_x^2 \langle x^2 \rangle_0, \quad \langle y^2 \rangle = \lambda_y^2 \langle y^2 \rangle_0, \quad \langle z^2 \rangle = \lambda_z^2 \langle z^2 \rangle_0 \tag{4.10}$$

Substituting Eq. (4.10) into Eq. (4.4) and using Eq. (4.8) leads to the elastic free energy of the affine network:

$$\Delta A_{\mathrm{el}} = \tfrac{1}{2} \nu k T (\lambda_x^2 + \lambda_y^2 + \lambda_z^2 - 3) \tag{4.11}$$

THE PHANTOM NETWORK MODEL

According to the *phantom network model* (James and Guth 1947), the junction points fluctuate over time without being hindered by the presence of the neighboring chains. The extent of the fluctuations is not affected by the macroscopic state of deformation. The term *phantom* derives from the assumed ability of the junctions to fluctuate in spite of their entanglements with network chains.

According to the theory, a small number of junctions are assumed to be fixed at the surface of the network, and the remaining ones are free to fluctuate over time. The instantaneous end-to-end vector of each chain may be represented as a sum of a mean $\bar{\mathbf{r}}_i$ and a fluctuation $\Delta \mathbf{r}_i$ from the mean:

$$\mathbf{r}_i = \bar{\mathbf{r}}_i + \Delta \mathbf{r}_i \tag{4.12}$$

The subscript i denotes that Eq. (4.12) is written for the ith chain.

The dot product of both sides of Eq. (4.12) is

$$r_i^2 = \bar{r}_i^2 + 2\bar{\mathbf{r}}_i \cdot \Delta \mathbf{r}_i + (\Delta r_i)^2 \tag{4.13}$$

Averaging both sides of Eq. (4.13) over all chains of the network in the state of rest and in the deformed state gives

$$\langle r^2 \rangle_0 = \langle \bar{r}^2 \rangle_0 + \langle (\Delta r)^2 \rangle_0$$
$$= \langle \bar{x}^2 \rangle_0 + \langle \bar{y}^2 \rangle_0 + \langle \bar{z}^2 \rangle_0 + \langle (\Delta x)^2 \rangle_0 + \langle (\Delta y)^2 \rangle_0 + \langle (\Delta z)^2 \rangle_0$$
$$\langle r^2 \rangle = \langle \bar{r}^2 \rangle + \langle (\Delta r)^2 \rangle$$
$$= \langle \bar{x}^2 \rangle + \langle \bar{y}^2 \rangle + \langle \bar{z}^2 \rangle + \langle (\Delta x)^2 \rangle + \langle (\Delta y)^2 \rangle + \langle (\Delta z)^2 \rangle \tag{4.14}$$

The average of the term $\bar{\mathbf{r}}_i \cdot \Delta \mathbf{r}_i$ in Eq. (4.13) is zero due to the fact that fluctuations of chain dimensions are uncorrelated with mean chain vectors.

At a given instant, the mean positions $\bar{\mathbf{r}}$ and fluctuations Δr of all chains exhibit distributions that may be assumed to be Gaussian (Flory 1976). The mean-squared values $\langle \bar{r}^2 \rangle_0$ and $\langle (\Delta r)^2 \rangle_0$ are related to $\langle r^2 \rangle$ according to theory by

$$\langle \bar{r}^2 \rangle_0 = \left(1 - \frac{2}{\phi} \right) \langle r^2 \rangle_0 \tag{4.15a}$$

$$\langle (\Delta r)^2 \rangle_0 = \frac{2}{\phi} \langle r^2 \rangle_0 \tag{4.15b}$$

Derivation of Eq. (4.15) is outlined in Appendix B.

The components of mean position $\bar{\mathbf{r}}$ of each chain transforms affinely with macroscopic deformation while fluctuations Δr are not affected, which results in the following relations:

$$\langle \bar{x}^2 \rangle = \lambda_x^2 \langle \bar{x}^2 \rangle_0, \quad \langle \bar{y}^2 \rangle = \lambda_y^2 \langle \bar{y}^2 \rangle_0, \quad \langle \bar{z}^2 \rangle = \lambda_z^2 \langle \bar{z}^2 \rangle_0$$
$$\langle (\Delta x)^2 \rangle = \langle (\Delta x)^2 \rangle_0, \quad \langle (\Delta y)^2 \rangle = \langle (\Delta y)^2 \rangle_0, \quad \langle (\Delta z)^2 \rangle = \langle (\Delta z)^2 \rangle_0 \tag{4.16}$$

Substituting Eq. (4.16) into Eq. (4.14) and using Eq. (4.15) and the condition of isotropy in the state of rest leads to

$$\langle r^2 \rangle = \left[\left(1 - \frac{2}{\phi} \right) \frac{\lambda_x^2 + \lambda_y^2 + \lambda_z^2}{3} + \frac{2}{\phi} \right] \langle r^2 \rangle_0 \qquad (4.17)$$

Using Eq. (3.2) and Eq. (4.17) in Eq. (4.4) leads to the following elastic free energy expression for the phantom network:

$$\Delta A_{el} = \tfrac{1}{2} \, \xi kT (\lambda_x^2 + \lambda_y^2 + \lambda_z^2 - 3) \qquad (4.18)$$

COMPARING THE TWO MODELS

The differences between the elastic free energy for an affine model [Eq. (4.11)] and the phantom model [Eq. (4.18)] arise specifically from the nature of transformations of chain dimensions built into the two models of the elementary theory.

The expression for the elastic free energy for both models may be represented as

$$\Delta A_{el} = \mathcal{F} kT (\lambda_x^2 + \lambda_y^2 + \lambda_z^2 - 3) \qquad (4.19)$$

where the *front factor* $\mathcal{F}$ equates to $\nu/2$ for the affine network model and to $\xi/2$ for the phantom network model. For a perfect tetrafunctional network, the front factor for the latter model is half the value for the former. Failure to recognize the differences in the assumptions of the two models led to a long series of heated arguments in the literature (Wall and Flory 1951; James and Guth 1953).

The simplified derivation of the elastic free energy for the Affine Network model presented in this chapter deviates from that obtained by Flory (1953) using more rigorous statistical mechanical analysis. According to Flory, the elastic free energy expression contains an additional logarithmic term that is a gaslike contribution resulting from the distribution of the cross-links over the sample volume. Thus the correct expression for the elastic free energy of the affine network model is

$$\Delta A_{el} = \frac{\nu kT}{2} (\lambda_x^2 + \lambda_y^2 + \lambda_z^2 - 3) - \mu kT \ln \left(\frac{V}{V_0} \right) \qquad (4.20)$$

where V is the final volume of the network. The presence of the logarithmic term is inconsequential for the force–deformation relations presented in Chapter 8. However, its presence is required in considering the equilibrium degree of swelling, as outlined in Chapter 7.

A more realistic theory, which subsumes the two limiting theories discussed here, is the subject of the following chapter.

5

STATISTICAL THEORY FOR REAL NETWORKS

As already mentioned, the basic challenge of the molecular theory of rubber elasticity is to relate the state of deformation at the molecular level to the externally applied macroscopic deformation. The two models described in the previous chapter, the affine network and phantom network models, are the two simplest models adopted for this purpose. In the affine network model, the junctions are assumed to be embedded securely in the network structure, showing no fluctuations over time as would be observed in a real network whose junctions exhibit rapid fluctuations about their mean positions. As a consequence of being embedded in the network, the junctions translate affinely with macroscopic strain. No assumption is made with regard to the parts of a chain between its junctions. The junctions in the phantom network model, on the other hand, reflect the full mobility of the chains subject only to the effects of the connectivity of the network. The position of each junction may be defined in terms of a time-averaged mean location and an instantaneous fluctuation from it. According to this other extreme case (James 1947), the mean locations of junctions transform affinely with macroscopic deformation, whereas the instantaneous fluctuations are not affected. The independence of the instantaneous fluctuations from the macroscopically applied state of deformation is a consequence of the phantomlike nature of the chains. During the course of their fluctuations the chains may pass freely through each other, being unaffected by the volume-exclusion effects of neighboring chains and therefore by the macroscopically applied deformation.

CONSTRAINED JUNCTION MODEL

A real network is expected to exhibit properties that fall between those of the affine and phantom network models, in that junction fluctuations do occur but not to the extent present at the phantom limit. This was first suggested by Ronca and Allegra (1975). A quantitative model of a network with fluctuations of junctions dependent nonaffinely on the macroscopic state of strain is given by the so-called *constrained junction* model of real networks (Flory 1977a). Detailed descriptions of its physical basis are given elsewhere (Flory 1977b; 1979, 1985a,b). According to this model, the fluctuations of junctions are affected by the copious interpenetration of their pendent chains with the spatially neighboring junctions and chains. The degree of interpenetration of a chain with its environment is of critical importance. This is described schematically for a tetrafunctional network in Figure 5.1a, where the four filled circles represent the junctions that are topological neighbors of a given junction (empty circle). The spatially neighboring junctions are shown by X's. The average number Γ of junctions within this domain is given by

$$\Gamma = \frac{4\pi}{3} \langle r^2 \rangle_0^{3/2} \frac{\mu}{V_0} \tag{5.1}$$

where μ/V_0 represents the number of junctions per unit volume in the reference state of the network. For typical networks, Γ is in the range 25–100. Unlike the limiting case of the phantom networks, a significant degree of rearrangement of junctions in the domain shown by the dashed circle is expected to occur upon macroscopic deformation of the network. This has the very important effect of rendering the junction fluctuations dependent on strain.

In the constrained junction network model, a given junction is assumed to be under the joint action of the phantom network and the constraint domains, as shown in Figure 5.1b. Point A locates the mean position of the junction in the phantom network. The large dashed circle of radius $\langle (\Delta R)^2 \rangle_{\text{ph}}^{1/2}$ represents the root-mean-square of the fluctuation domain for the junction in the phantom network. Point B locates the mean location of constraints, which is at a distance $\bar{s}$ from the phantom center. The small dashed circle of radius $\langle (\Delta s)^2 \rangle_0^{1/2}$ represents the root-mean-square size of the constraint domain in which the junction would fluctuate under the effect of constraints only. Point C locates the mean position of the junction under the combined action of the phantom network and constraints. Point D denotes the instantaneous location of the junction at a dis-

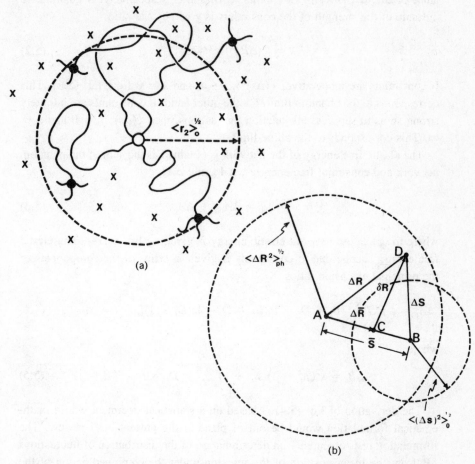

Figure 5.1 (a) Tetrafunctional junction (empty circle) surrounded by spatially neighboring junctions (×'s) and four topological junctions (filled circles). The dashed line indicates the radius of the average volume pervaded by a network chain. (b) Various variables defining the mean and instantaneous positions of a given junction. Point A is the mean position of the junction in the phantom network. Point B is the center of entanglements, located at a distance $\bar{s}$ from point A. Point D is the instantaneous location of the junction in the real network, located at a distance ΔR from the phantom center and ΔS from the constraint center. The large dashed circle of radius $\langle (\Delta R)^2 \rangle_{ph}^{1/2}$ denotes the average region in which the junction would fluctuate in the absence of constraints. The small dashed circle of radius $\langle (\Delta s)^2 \rangle_0^{1/2}$ denotes the region in which the junction would fluctuate under the action of constraints only. Point C is the center about which the junction fluctuates under the combined effects of the phantom network and the constraints.

tance of ΔR, Δs, and δR from points A, B, and C, respectively. A quantitative measure of the strength of the constraints is given by the ratio

$$\kappa = \langle (\Delta R)^2 \rangle_{\text{ph}} / \langle (\Delta s)^2 \rangle_0 \tag{5.2}$$

If constraints are inoperative, $\langle (\Delta s)^2 \rangle_0 \rightarrow \infty$ and $\kappa = 0$ from Eq. (5.2). This corresponds to the phantom limit. On the other hand, if constraints are infinitely strong so as to suppress all junction fluctuations, then $\langle (\Delta s)^2 \rangle_0 = 0$ and $\kappa \rightarrow \infty$. This corresponds to the affine limit.

The elastic free energy of the network is obtained as the sum of the phantom network and constraint free energies, ΔA_{ph} and ΔA_c:

$$\Delta A_{\text{el}} = \Delta A_{\text{ph}} + \Delta A_c \tag{5.3}$$

where the phantom network elastic energy is given by Eq. (4.18). The elastic free energy change due to constraints is given in terms of the components of the principal extension ratios:

$$\Delta A_c = \tfrac{1}{2} \mu kT \sum_t [B_t + D_t - \ln(B_t + 1) - \ln(D_t + 1)], \qquad t = x, y, z \tag{5.4}$$

where

$$B_t = \kappa^2 (\lambda_t^2 - 1)(\lambda_t^2 + \kappa)^{-2}, \qquad D_t = \lambda_t^2 \kappa^{-1} B_t \tag{5.5}$$

The derivation of Eq. (5.4) is based on a statistical treatment whose mathematical formulation would be out of place in the present brief review. The formulation rests essentially on determination of the distribution of fluctuations δR from the mean position of the junction under the combined action of the phantom network and constraint effects in the deformed network. The reader is referred to the original paper (Flory 1977a) for the mathematical derivation of the additional elastic free energy ΔA_c due to the action of constraints on δR. For $\kappa = 0$, $\Delta A_c = 0$ and the elastic free energy of the network is equal to that of the phantom network. As κ increases indefinitely, the elastic free energy converges to that of the affine network given by Eq. (4.20). Thus the constrained junction model represents a network with elastic free energy intermediate in value between the phantom and the affine network limits. As will be described in more detail in Chapter 8, the behavior of the real network is closer to that of the affine network model at low deformations and approaches the phantom network model as the deformation increases.

The κ parameter of the constrained junction model, defined by Eq. (5.2), can be interpreted in terms of the molecular constitution of the network (Erman and Flory 1982) by assuming it to be proportional to the average number of junctions in the domain occupied by a network chain [see Eq. (5.1)]. Thus

$$\kappa = I(N_A d/2)^{3/2}(\langle r^2 \rangle_0/M)^{3/2}(\xi/V_0)^{-1/2}$$
$$= I(2/\phi)(N_A d)(\langle r^2 \rangle_0/M)^{3/2}M_c^{1/2} \qquad (5.6)$$

where I is the constant of proportionality, N_A is Avogadro's number, d is the network density, and M is the molecular weight of a chain with end-to-end mean-square length $\langle r^2 \rangle_0$. Equation (5.6) indicates that κ is inversely proportional to the square root of cycle rank density ξ/V_0, inversely proportional to the functionality ϕ, and directly proportional to the square root of network chain molecular weight M_c.

SLIP-LINK MODEL

The theory of rubber elasticity based on the phantom, affine, and constrained junction models focuses attention on the junctions. Factors that may be affecting points along the chain contours do not appear explicitly in the formulation. An alternative model by Edwards and colleagues (Ball et al. 1981), referred to as the *slip-link model*, incorporates the effects of entanglements along the chain contour into the elastic free energy. According to this model, a link joins two different chains as shown in Figure 5.2. The link may slide a distance along the

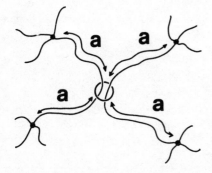

Figure 5.2 Slip-link joining two chains. The slip-link is assumed to slide a distance a along the chains.

contour length of the chains. The actions of the slip-links are held to be equivalent to additional cross-links in the network, but not of course with a one-to-
one correspondence. The statistical derivation of the free energy is based on a
highly sophisticated mathematical model, referred to as the *replica model*. The
resulting expression is

$$\Delta A_{el} = \frac{1}{2} N_c kT \left\{ \sum_{i=1}^{3} \lambda_i^2 + \frac{N_s}{N_c} \sum_{i=1}^{3} \left[\frac{(1 + \eta)\lambda_i^2}{1 + \eta\lambda_i^2} + \log(1 + \eta\lambda_i^2) \right] \right\} \quad (5.7)$$

where N_c and N_s are the number of chemical cross-links and slip-links, respectively, and $\eta = 0.2343 \ldots$ The first term on the right-hand side of Eq. (5.7)
is the elastic energy of the phantom network. The second term denotes contributions to ΔA_{el} from slip-links and corresponds to ΔA_c of Eq. (5.3). There is
no restriction on the number of slip-links N_s. Thus the ratio N_s/N_c may be quite
large, and the elastic free energy of the network with slip-links may exceed that
of the affine network. This differs significantly from the predictions of the constrained junction model, according to which the upper limit of ΔA_{el} is that for
an affine network.

OTHER MODELS

There are still other molecular approaches to a general theory, for example,
those based on a tube model (Gaylord 1982) and a van der Waals model (Kilian
et al. 1986; Boué and Vilgis 1986).

6

ELASTIC EQUATIONS OF STATE

INTRODUCTION

An equation of state is an equation interrelating the various properties required to characterize a system. The elastic equation of state for a rubber network thus specifies the relationship between the applied forces, the resulting deformations, and the molecular structure of the network. It is obtained according to the thermodynamic expression

$$\tau_t = V^{-1}\lambda_t(\partial\Delta A_{el}/\partial\lambda_t)_{T,V}, \qquad t = 1, 2, 3 \tag{6.1}$$

where τ_t is the stress along the tth coordinate direction, and is defined as the force per unit deformed area (Flory 1961). The quantity λ_t is the ratio of the final length to the reference length along that direction. The subscripts T,V indicate that the differentiation is performed at fixed temperature and volume. The index t may take values 1, 2, 3, as indicated in Eq. (6.1), or x, y, z, as given in Chapter 4. The two sets, denoting the three principal coordinate directions, are used interchangeably throughout the text.

Equation (6.1) may conveniently be interpreted by considering a prismatic block of a network with sides L_{01}, L_{02}, and L_{03}, and volume V_0 in the reference state (Fig. 6.1a). In the case of network formation in solution, V_0 represents the total volume of polymer and solvent. Figure 6.1b shows the dimensions L_{i1},

41

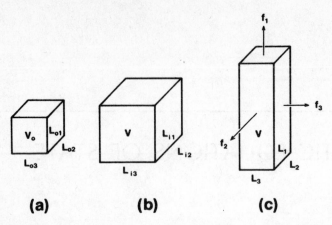

(a)   **(b)**   **(c)**

Figure 6.1 (a) A prismatic block at the reference volume V_0. The reference state is the state at which the network is formed. (b) The same block at the initial state before the application of the deformation. The volume V may be different from V_0 due to a difference in amount of solvent present or in temperature. (c) The final dimensions of the block after the stresses are applied. Equivalence of volumes at the initial and final states indicates that the volume of the network remains constant and that no solvent enters or leaves the network during deformation under the stresses.

L_{i2}, and L_{i3} of the network at the start of the experiment, where the volume V may be different from V_0 depending on the amount of solvent present relative to that during formation. Figure 6.1c depicts the deformed state under forces f_1, f_2, and f_3 applied along the three coordinate directions. The equality of the volumes before and after the application of the stresses, as indicated in Figures 6.1b and 6.1c, is the result of assuming incompressibility. As mentioned in Chapter 1, experimental data in simple tension show that the volume change in ordinary networks is negligibly small in comparison to changes in linear dimensions.

The deformation ratio λ_t along the direction t is defined as

$$\lambda_t = L_t/L_{0t}, \qquad t = 1, 2, 3 \tag{6.2}$$

Thus three deformation ratios are related to the final volume and reference volume by

$$\lambda_1\lambda_2\lambda_3 = V/V_0 \tag{6.3}$$

Inasmuch as V_0 and V are fixed quantities in a given experiment, Eq. (6.3) states that only two of the three λ_t terms are independent. The deformation ratio α_t relative to the initial dimensions is defined by

$$\alpha_t = L_t/L_{it} = (V/V_0)^{-1/3}\lambda_t, \qquad t = 1, 2, 3 \qquad (6.4)$$

The third term of Eq. (6.4) follows from the assumption that the network is isotropic in the undistorted state.

A knowledge of the volume fraction of polymer present during network formation and during the stress–strain experiment is essential for any interpretation by the molecular theory in terms of equations of state. The two volume fractions defined for this purpose are

$$v_{2C} = V_d/V_0 \qquad (6.5)$$

$$v_2 = V_d/V_i \qquad (6.6)$$

where V_d is the volume of the dry network, and V_i is the volume of the network plus solvent at the start of the stress–strain experiment. Thus the conversion factor $(V/V_0)^{-1/3}$ in Eq. (6.4) can be alternatively written as $(v_{2C}/v_2)^{1/3}$.

The deformation ratios λ_1, λ_2, and λ_3 are not independent of one another due to the requirement indicated by Eq. (6.3). It is therefore necessary to rewrite Eq. (6.1) in the form

$$\tau_t = V^{-1}\lambda_t \sum_{i=1}^{3} (\partial\Delta A_{el}/\partial\lambda_i^2)\partial\lambda_i^2/\partial\lambda_t \qquad (6.7)$$

Using the relation $\Delta A_{el} = (\mathfrak{F}kT)(\lambda_1^2 + \lambda_2^2 + \lambda_3^2 - 3)$ from Chapter 4, Eq. (6.7) becomes

$$\tau_t = \left(\frac{\mathfrak{F}kT}{V}\right)\lambda_t \sum_{i=1}^{3} \frac{\partial\lambda_i^2}{\partial\lambda_t} \qquad (6.8)$$

An alternative derivation of the stresses is possible by the use of the so-called *Treloar relations* involving differences in stresses (Treloar 1975).

Specification of the λ_t in Eq. (6.8) for any type of deformation then gives the desired corresponding equation of state. Setting $\mathfrak{F} = \nu/2$ yields the affine limit and $\mathfrak{F} = \xi/2$ the phantom limit. For a perfect network, $\xi = \nu(1 - 2/\phi)$.

Derivations of the equations of state for the constrained junction model are presented in Appendix C.

UNIAXIAL EXTENSION (OR COMPRESSION)

The state of deformation for extension or compression along the x axis is

$$\lambda_1 = \alpha(V/V_0)^{1/3} \tag{6.9a}$$

$$\lambda_2 = \lambda_3 = \alpha^{-1/2}(V/V_0)^{1/3} \tag{6.9b}$$

where α is the ratio of the final length along the direction of stretch to the initial undistorted length at volume V. For compression, $\alpha < 1$.

Writing Eq. (6.8) as

$$\tau_1 = \frac{\mathfrak{F}kT}{V}\lambda_1 \sum_{i=1}^{3} \frac{\partial \lambda_i^2}{\partial \alpha}\frac{\partial \alpha}{\partial \lambda_1} \tag{6.10}$$

and using Eqs. (6.9) leads to

$$\tau_1 = 2\left(\frac{\mathfrak{F}kT}{V}\right)\left(\frac{V}{V_0}\right)^{2/3}(\alpha^2 - \alpha^{-1}) \tag{6.11}$$

The force f acting along the x axis is obtained by multiplying both sides of Eq. (6.11) by the deformed area:

$$f = 2\left(\frac{\mathfrak{F}kT}{L_{i1}}\right)\left(\frac{V}{V_0}\right)^{2/3}(\alpha - \alpha^{-2}) \tag{6.12}$$

Experimental data are generally represented in terms of the *reduced stress* $[f^*]$ defined as

$$[f^*] \equiv \frac{fv_2^{1/3}}{A_d(\alpha - \alpha^{-2})} \tag{6.13}$$

where A_d is the cross-sectional area of the undeformed, dry sample. Using Eq. (6.12) in Eq. (6.13) leads to

$$[f^*] = 2\left(\frac{\mathfrak{F}kT}{V_d}\right) v_{2C}^{2/3} \tag{6.14}$$

The use of this equation is illustrated by the first of the two problems presented and solved in Appendix E.

The reduced stress may be interpreted as the shear modulus of a network. Equation (6.14) shows that for a phantom or affine network, the modulus or reduced stress is predicted to be independent of deformation. Experiments show, however, that it depends rather strongly on deformation (Treloar 1975; Mark 1975). This dependence of $[f^*]$ on α is well predicted by the constrained junction model. Typical experimental data are presented and compared with theoretical predictions in Chapter 8. A Fortran program for calculating $[f^*]$ for this model is given in Appendix D.

The applied force f in uniaxial extension has two effects on the network at the molecular level. Part of the force is used in changing the configurations of the network chains to a less disordered state. This component changes the entropy of the network only, and is referred to as the entropic component. The remaining part of the force is used in changing the energy but not the entropy of each chain, simply by forcing the chains to assume configurations of different energy. This component, denoted by f_e, is referred to as the energetic contribution to the total force (Flory et al. 1960). From thermodynamics, we have

$$\frac{f_e}{f} = -T\left[\frac{\partial \ln (f/T)}{\partial T}\right]_{L,V} \tag{6.15}$$

where the subscripts L,V denote that differentiation is performed at constant length and volume. Equation (6.15) gives rise to a very simple general relationship between f_e/f and the chain dimensions that is most easily derived in the case of a phantom network. Dividing the right-hand side of Eq. (6.12) by T, taking the logarithm, and differentiating with respect to T at constant L and V (and therefore at constant α) leads to the simple result

$$\frac{f_e}{f} = \frac{T\,d \ln \langle r^2 \rangle_0}{dT} \tag{6.16}$$

where the relation $d \ln \langle r^2 \rangle_0/dT = \frac{2}{3} d \ln V_0/dT$ is used. The term $d \ln \langle r^2 \rangle_0/dT$ is a molecular parameter referred to as the temperature coefficient of the unperturbed mean-squared end-to-end vector (Flory 1969).

Experiments are commonly carried out at constant pressure rather than at constant volume. In this case, further correction terms are added to Eq. (6.16). Detailed discussion of such terms and experimental analysis are given in Chapter 9.

BIAXIAL EXTENSION

The state of deformation for biaxial extension, where the dimensions of the network are changed independently along axes 1 and 2, is given by

$$\lambda_1 = \alpha_1 (V/V_0)^{1/3}$$
$$\lambda_2 = \alpha_2 (V/V_0)^{1/3} \tag{6.17}$$
$$\lambda_3 = (\alpha_1\alpha_2)^{-1} (V/V_0)^{1/3}$$

where α_1 and α_2 denote the ratios of final to undistorted initial lengths along directions 1 and 2, respectively. Substitution of Eqs. (6.17) into Eq. (6.8) leads to

$$\tau_1 = 2\left(\frac{\mathcal{F}kT}{V}\right)\left(\frac{V}{V_0}\right)^{2/3}\left(\alpha_1^2 - \frac{1}{\alpha_1^2\alpha_2^2}\right)$$
$$\tau_2 = 2\left(\frac{\mathcal{F}kT}{V}\right)\left(\frac{V}{V_0}\right)^{2/3}\left(\alpha_2^2 - \frac{1}{\alpha_1^2\alpha_2^2}\right) \tag{6.18}$$

For equibiaxial extension, such as obtained in the inflation of a spherical balloon, $\alpha_1 = \alpha_2 = \alpha$. Therefore, from Eq. (6.18),

$$\tau = 2\left(\frac{\mathcal{F}kT}{V}\right)\left(\frac{V}{V_0}\right)^{2/3}(\alpha^2 - \alpha^{-4}) \tag{6.19}$$

Results of biaxial extension experiments are compared with predictions of theory in Chapter 8.

PURE SHEAR

The state of pure shear is essentially a biaxial loading under the stresses τ_1 and τ_2 such that there is no change in length along the 2 direction. Thus, the state of deformation is described by Eq. (6.17) with $\alpha_2 = 1$. Letting $\alpha_1 = \alpha$ and $\alpha_2 = 1$ in Eq. (6.18), the stresses obtained are

$$\tau_1 = 2\left(\frac{\mathfrak{F}kT}{V}\right)\left(\frac{V}{V_0}\right)^{2/3}(\alpha^2 - 1/\alpha^2)$$

$$\tau_2 = 2\left(\frac{\mathfrak{F}kT}{V}\right)\left(\frac{V}{V_0}\right)^{2/3}(1 - 1/\alpha^2)$$

$$(6.20)$$

Results of pure shear experiments presented in Chapter 8 show significant deviations from the predictions of Eq. (6.20) obtained according to either the phantom network or the affine network models. The experimental data, however, are in satisfactory agreement with predictions of the constrained junction theory, as derived in Appendix C.

7

SWELLING OF NETWORKS

All synthetic and biological networks swell when exposed to low molecular weight solvents. The degree of swelling at equilibrium depends on factors such as temperature, length of the network chains, size of the solvent molecules, and the strength of thermodynamic interaction between the polymer chains and solvent molecules.

As in the previous applications, the thermodynamics of the system may be described by the change in the Gibbs free energy ΔG of the system. The pressure–volume product does not change significantly in swelling, however, and thus ΔG can be replaced by the change in Helmholtz free energy ΔA. The total change results from the change in the elastic free energy ΔA_{el} of the network upon isotropic dilation with the introduction of the solvent, and from the change in the free energy of mixing ΔA_{mix} of the solvent molecules with the chains constituting the network. It is assumed (Flory 1953; Treloar 1975; Queslel and Mark 1985a) that the change in the total free energy is the direct sum of ΔA_{el} and ΔA_{mix}. Thus

$$\Delta A = \Delta A_{el} + \Delta A_{mix} \tag{7.1}$$

Expressions for ΔA_{el} for a phantom network and an affine network are given by Eqs. (4.18) and (4.20), respectively. The deformation ratios λ_1, λ_2, and λ_3

in these equations must now correspond to the state of isotropic dilation, that is,

$$\lambda_1 = \lambda_2 = \lambda_3 = (V_m/V_0)^{1/3} \equiv (v_{2C}/v_{2m})^{1/3} \tag{7.2}$$

where V_m is the volume of solvent plus polymer, and v_{2m} is the volume fraction of polymer at equilibrium (maximum) degree of swelling when exposed to excess solvent. Substituting Eq. (7.2) into Eqs. (4.20) and (4.18), respectively, leads to

$$\Delta A_{el} = \frac{3\nu kT}{2}\left[\left(\frac{v_{2C}}{v_{2m}}\right)^{2/3} - 1\right] - \mu kT \ln\frac{v_{2C}}{v_{2m}} \quad \text{(affine)} \tag{7.3}$$

and

$$\Delta A_{el} = \frac{3\xi kT}{2}\left[\left(\frac{v_{2C}}{v_{2m}}\right)^{2/3} - 1\right] \quad \text{(phantom)} \tag{7.4}$$

The second term of Eq. (7.1), ΔA_{mix}, for mixing polymer chains with solvent is given by the Flory–Huggins relationship (Flory 1942, 1953; Huggins 1942):

$$\Delta A_{mix} = kT(n_1 \ln v_1 + n_2 \ln v_2 + \chi n_1 v_2) \tag{7.5}$$

where n_1 and n_2 are the numbers of solvent and polymer molecules, respectively. For a cross-linked network, $n_2 = 1$. The quantity χ is the interaction parameter for the polymer–solvent system (Flory 1953).

Introduction of solvent molecules into the system results in (1) an increase in ΔA_{el} due to the decrease of entropy of the network chains upon dilation, and (2) a decrease in ΔA_{mix} primarily due to the increase in the entropy of mixing the solvent molecules with the network chains. A state of equilibrium swelling is obtained when the two changes balance each other. Mathematically this state is expressed as

$$\left(\frac{\partial \Delta A}{\partial n_1}\right)_{T,p} = \left(\frac{\partial \Delta A_{mix}}{\partial n_1}\right)_{T,p} + \left(\frac{\partial \Delta A_{el}}{\partial n_1}\right)_{T,p} = 0 \tag{7.6}$$

where the subscripts T, p indicate that the differentiations are made at constant temperature and pressure. Performing the differentiations indicated in Eq. (7.6) (Queslel and Mark 1985a, 1988) one obtains for an affine network

$$\ln(1 - v_{2m}) + \chi v_{2m}^2 + v_{2m} + B\left(1 - \frac{2}{\phi}\right)\left[\left(\frac{v_{2m}}{v_{2C}}\right)^{1/3} - \frac{\mu}{\nu}\frac{v_{2m}}{v_{2C}}\right] = 0 \quad (7.7)$$

and for a phantom network,

$$\ln(1 - v_{2m}) + \chi v_{2m}^2 + v_{2m} + B(v_{2m}/v_{2C})^{1/3} = 0 \qquad (7.8)$$

where

$$B = (V_1/RT)(\xi kT/V_0) \qquad (7.9)$$

In the highly swollen state, the constrained junction theory (Flory 1977a) indicates that a real network exhibits properties closer to those of the phantom network model. Consequently Eq. (7.8) is a more realistic representation for equilibrium swelling.

Equation (7.8) may thus be used to estimate the average chain length M_c between cross-links. Using the relation $\xi kT/V_0 = (1 - 2/\phi)\rho RT/M_c$ obtained from Eq. (3.3) and solving Eq. (7.8) for M_c leads to

$$M_c = -\frac{\rho(1 - 2/\phi)V_1 v_{2C}^{2/3}\,v_{2m}^{1/3}}{\ln(1 - v_{2m}) + \chi v_{2m}^2 + v_{2m}} \qquad (7.10)$$

Characterization of the network chain length in this manner requires precise measurement of v_{2m}. The value of χ at this value of v_{2m} is also required. This application is illustrated by the second of the two problems presented and solved in Appendix E. Alternatively, Eq. (7.8) may serve as a convenient means of evaluating the χ parameter. Solving Eq. (7.8) for χ, one obtains

$$\chi = -\frac{\ln(1 - v_{2m}) + v_{2m} + Bv_{2C}^{2/3}\,v_{2m}^{1/3}}{v_{2m}^2} \qquad (7.11)$$

The value of the cross-link density as embodied in the factor B and the corresponding value of v_{2m} are required for evaluating χ from Eq. (7.11). Use of several network samples with different cross-link densities leads of course to an estimate of χ as a function of v_{2m} (Bahar et al. 1987).

Swelling is also frequently used as a preliminary to elongation measurements. In this application, a large-enough quantity of a solvent, generally nonvolatile, is introduced into the network to facilitate the approach to elastic equilibrium or to suppress crystallization. This also has the advantageous effect of decreasing the extent to which the reduced stress changes with elongation (Treloar 1975; Mark 1975), as is described in Chapter 8. There is, however, a complication that can occur in the case of networks of polar polymers at relatively high degrees of swelling. The observation is that different solvents at the same degree of swelling can have significantly different effects on the elastic force (Yu and Mark 1974). This is apparently due to a *specific solvent effect* on the unperturbed dimensions or on $V_0 \propto \langle r^2 \rangle_0^{3/2}$, which appear in the basic relationships given in Chapters 4 and 5. Although frequently observed in studies of the solution properties of uncross-linked polymers, the effect is not yet well understood (Dondos and Benoit 1971). It is apparently partly due to the effect of the solvent's dielectric constant on the coulombic interactions between parts of a chain, but probably also to solvent–polymer segment interactions that change the conformational preferences of the chain backbone.

8

FORCE AS A FUNCTION OF DEFORMATION

INTRODUCTION

Force–deformation relations for polymeric networks are generally obtained by mechanical experiments where a polymeric network is subjected to a specified state of deformation and the forces required to maintain this state are measured. A typical apparatus was described in Chapter 1. As already mentioned, it is only the equilibrium state of the deformed network that is of interest for the statistical theory. Thus sufficient time is assumed to pass between the application of the deformations and the final time of measurements beyond which no appreciable changes in the forces take place. Alternatively, a state of constant stress may be imposed on a network, and the resulting deformations may be measured at equilibrium. In either case, the relationship of the stresses to components of the deformation tensor are given by Eq. (6.1). In this chapter, force–deformation relations predicted by the elementary molecular theory and the constrained junction theory are compared with experimentally observed data.

UNIAXIAL EXTENSION OR COMPRESSION

The reduced stress $[f^*]$ given by Eq. (6.13) is independent of deformation in both the phantom and the affine network models. In Figure 8.1 the reduced

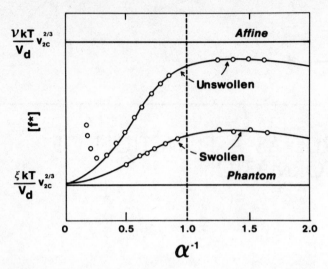

Figure 8.1   Reduced stress as a function of reciprocal extension ratio. The upper and lower horizontal lines represent results from affine and phantom network models, respectively. Circles show representative data from experiments, and the curves are from the constrained junction theory.

force is shown as a function of reciprocal α by the two horizontal lines for the affine network model (with any functionality ϕ) and a phantom network model (with specific but arbitrary ϕ), respectively. The advantage of this type of representation is that the data in elongation can be well represented by a straight line over a wide range in α^{-1}. Thus, elongation data have very frequently been fitted to the relationship (Mooney 1948; Rivlin 1948)

$$[f^*] = 2C_1 + 2C_2\alpha^{-1} \tag{8.1}$$

where $2C_1$ and $2C_2$ are constants independent of α (Treloar 1975; Mark 1975). Thus $2C_1$ has been taken as an estimate of the high-deformation modulus (the phantom network model limit). Similarly, $2C_1 + 2C_2$ has been used as an estimate of the low-deformation modulus (the affine network model limit). Correspondingly, $2C_2$ and its normalized value $2C_2/2C_1$ can represent a measure of the extent to which the deformation becomes nonaffine as α increases. The circles represent typical results of experiments (Erman and Flory 1978; Pak and Flory 1979; Mark 1979a,b). The upper set of points denotes results on networks deformed in the unswollen state. Data points are approximately independent of deformation in the compression region ($\alpha^{-1} > 1$), whereas a steady decrease is observed in tension ($\alpha^{-1} < 1$), accompanied by a sharp increase at low values of α^{-1}. Thus, real networks exhibit reduced forces ranging in values from those close to the affine network model limit in compression and at low extensions, to those closer to the phantom network model limit at higher extensions. The upturn at high extensions result from either strain-induced crystallization or finite chain extensibility, as discussed in further detail in Chapter 12.

The lower set of circles shows typical results of experiments on networks deformed in the swollen state. Experiments indicate that the variation of $[f^*]$ with α^{-1} becomes weaker, in general, with increasing degree of swelling. Also, if the upturn in $[f^*]$ were due to strain-induced crystallization, it would be diminished or entirely suppressed by the melting-point depression caused by the solvent (Flory 1953; Mark 1979a,b). At high degrees of swelling ($v_2 \sim 0.2$) the data points coincide practically with the phantom network model limit over the entire range of α^{-1}. Similarly, dependence of $[f^*]$ on α weakens in networks prepared in solution and studied dried, as is described in Chapter 11.

The curves through the representative data points indicate predictions of the constrained junction theory (Erman and Flory 1982), for which the expression for $[f^*]$ is given by Eq. (C.6). The theory has been successful in predicting data in tension and compression at different degrees of swelling. According to the molecular model on which the constrained junction theory is based, the constraints are more pronounced at low deformations. This renders the transformations of junction positions strongly dependent on macroscopic deformation. The behavior is therefore closer to that of the affine network at low deformations. At high extensions or swelling, the effects of constraints diminish and the network approaches the phantom network model.

BIAXIAL EXTENSION

A state of biaxial extension is obtained by stretching a thin sheet of elastomer in two directions in its own plane. The state of deformation is represented by Eqs. (6.17), where directions 1 and 2 are in the plane of the sheet and direction 3 is perpendicular to it. The extension ratios α_1 and α_2 may be varied independently. Although experimental analysis of the network in biaxial extension is more complicated than in simple uniaxial extension, results verify the validity of the statistical theory, as is shown in Figure 8.2. The circles in the figure show results of measurements by Kawai et al. (Obata et al. 1970), on unswollen

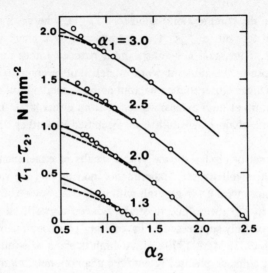

Figure 8.2 Difference of true stresses, $\tau_1 - \tau_2$, in biaxial extension shown as a function of extension ratio α_2. Circles represent data from Obata et al. (1970). Dashed and solid curves are calculated according to the elementary molecular and constrained junction theories, respectively.

rubber sheets by keeping the α_1 values constant and modifying the α_2 values. The ordinate values represent the stress difference $\tau_1 - \tau_2$. The dashed curves are calculated from the elementary molecular theory using Eqs. (6.18), and the solid curves are obtained from the constrained junction theory [Eq. (C.7)]. The improvement provided by the latter theory at lower extensions is readily seen.

PURE SHEAR

Results of experiments on natural rubber in pure shear are shown by the circles in Figure 8.3 (Rivlin and Saunders 1951). The data were obtained by keeping α_2 at unity and modifying $\alpha_1 = \alpha$, and α_3 such that $\alpha_3 = 1/\alpha$. The ordinate values represent the ratio $\tau_1/(\alpha^2 - \alpha^{-2})$, where τ_1 is the true stress applied along direction 1. The horizontal dashed line represents the prediction of the elementary molecular theory obtained from to Eq. (6.20). The solid curve is obtained (Erman 1981) by using the constrained junction theory represented by

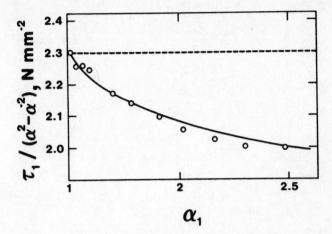

Figure 8.3 Comparison of experiment and theory for pure shear. The sample is fixed in the x_2 direction, and τ_1 and α_1 represent the stress and extension ratio along the x_1 direction, respectively. The circles represent data from Rivlin and Saunders (1951). The horizontal dashed line is obtained from the elementary molecular theory, and the solid curve, which falls very close to the experimental points, is from the constrained junction theory.

Eq. (C.8) for the case of pure shear. The dependence calculated from the elementary molecular theory is seen to be totally at variance with experiment, but that from the constrained junction theory gives excellent agreement.

9

FORCE AS A FUNCTION OF TEMPERATURE

INTRODUCTION

As was illustrated in Chapter 4, one of the most important thermodynamic quantities is the free energy. It is a state function and has been given this particular name because it represents that portion of the energy available ("free") to do work under specified conditions (Atkins 1982). The type of free energy used in Chapter 4 is called the *Helmholtz free energy* $A \equiv E - TS$, and is most useful under conditions of constant temperature and constant volume. (The fact that the theory proceeds through the Helmholtz free energy complicates things for experimentalists. They must either do the experiments at constant volume, which is very difficult, or correct their constant-pressure data to constant volume, which requires model-based approximations.)

The second type of free energy of interest to physical chemists is the *Gibbs free energy*, $G \equiv H - TS$. It is more convenient for analysis of systems at constant temperature and constant pressure. For such a process G must decrease, consistent with nature's attempt to decrease the energy of a system while simultaneously increasing its entropy, or disorder.

Its relevance to rubberlike elasticity can be illustrated by analysis of force–temperature (*thermoelastic*) measurements (Flory et al. 1960). Such experiments, first described qualitatively in Chapter 1, have now been carried out quantitatively for a wide variety of elastomers. The basic question in this analysis was raised in a preliminary manner in Chapter 6. It concerns the nature of the free energy increase attending the deformation of an elastomer, or the free energy decrease accompanying its spontaneous retraction when the deforming stress is removed. Specifically, how much of the free energy change is due to changes in energy and how much to changes in entropy? If the experiments are carried out at constant V, ΔA_{el} is of greatest interest; experiments at constant p, however, shift the focus to ΔG_{el}, that is, to the change in the Gibbs elastic free energy. The usual analysis, summarized below, pertains directly to experiments at constant V, and this part is based entirely on purely thermodynamic arguments. Subsequent adoption of a molecular model then permits modification of these equations to make them applicable to the much simpler experiments carried out at constant pressure.

As was mentioned in Chapter 1, the pressure exhibited by an ideal gas is entirely due to entropy effects, specifically the tendency for the gas to increase its entropy by increasing its volume (Atkins 1982). This absence of energy effects results from the fact that the gas molecules are assumed not to repel one another through an excluded-volume effect, nor to attract one another through the usual nonbonded intermolecular interactions. One repercussion is the prediction, from $pV = nRT$, that, at constant volume, the pressure should be directly proportional to the absolute temperature.

An ideal elastomer can be analogously defined as one in which the elastic force is entirely due to entropy effects, specifically, the tendency of the network chains to increase their entropy by retracting to more random conformations. This ideality would occur when (1) intermolecular interactions do not depend on degree of deformation (one of the major assumptions of the molecular theories), (2) the dependence of intermolecular interactions on the spacing between chains is nullifed by constraining measurements to constant volume, and (3) the conformational energies of the network chains do not change with deformation [which makes the unperturbed dimensions $\langle r^2 \rangle_0 \propto V_0^{2/3}$ in Eq. (6.11) independent of temperature]. As can be seen from this equation, the force or stress at constant elongation is then proportional to absolute temperature.

THEORY

The described proportionality between f^* and T is generally not observed, since different conformations of a polymer chain generally do *not* correspond to the same energy. The extent of this thermodynamic nonideality can be represented by the energetic contribution to the total elastic force (Flory et al. 1960; Treloar

1975; Mark 1973, 1976; Godovsky 1986). In the case of uniaxial deformation (elongation or compression), it is given by

$$f_e \equiv (\partial E/\partial L)_{T,V} \tag{9.1}$$

The most useful form for representing such information is the fraction f_e/f of the total force that is of energetic origin. Purely thermodynamic analysis, described in Chapter 6, shows this ratio to be given by the expression

$$\frac{f_e}{f} = -T\left[\frac{\partial \ln(f/T)}{\partial T}\right]_{L,V} \tag{6.15}$$

and it is thus seen to be immediately accessible from force–temperature measurements at constant length and volume. Some thermoelastic studies have indeed been carried out at constant volume (Allen et al. 1971), as required by this thermodynamically exact equation but, because of severe experimental difficulties encountered in meeting this requirement, most experiments are conducted at constant pressure. Transformation of Eq. (6.15) to a form suitable for analysis of such isobaric data requires recourse to an appropriate elastic equation of state. Adoption of the equation of state resulting from the simplest molecular or statistical theories of polymer networks for this purpose gives the result (Flory 1961)

$$\frac{f_e}{f} = -T\left[\frac{\partial \ln(f/T)}{\partial T}\right]_{L,p} - \frac{\beta T}{(\alpha^3 - 1)} \tag{9.2}$$

for measurements at constant length and pressure, where $\beta = (1/V)(\partial V/\partial T)_p$ is the thermal expansion coefficient of the network. Similarly,

$$\frac{f_e}{f} = -T\left[\frac{\partial \ln(f/T)}{\partial T}\right]_{\alpha,p} + \frac{\beta T}{3} \tag{9.3}$$

for measurements at constant pressure and deformation.

The above equations are readily modified for other types of strain. For example, a network in torsion is characterized by the torsion couple M and angle of torsion ϕ (which replace, respectively, the quantities f and L used to characterize a network in uniaxial deformation) (Treloar 1975). For this type of deformation the basic thermoelastic equations are

$$M_e = (\partial E / \partial \phi)_{T,V,L} \tag{9.4}$$

$$\frac{M_e}{M} = -T \left[\frac{\partial \ln(M/T)}{\partial T} \right]_{L,V,\phi} \tag{9.5}$$

$$= -T \left[\frac{\partial \ln(M/T)}{\partial T} \right]_{L,p,\phi} + \beta T \tag{9.6}$$

which parallel Eq. (9.1), (6.15), and (9.2), respectively.

As described above, the energy changes may be assumed to be intramolecular, resulting from changes in conformational energies of the network chains. [Thus, since the nonideality is intramolecular, it cannot be removed by diluting the chains (swelling the network) nor by increasing the lengths of the chains (decreasing the degree of cross-linking). In this respect, elastomers are different from gases, which can be made to behave ideally by decreasing the pressure to a sufficiently low value (Atkins 1982).] Because f_e is intramolecular, the same theoretical equation of state mentioned above may also be used to obtain a molecular interpretation of thermoelastic data and the quantity f_e/f derived therefrom. The result is given by the equation, derived in Chapter 6,

$$\frac{f_e}{f} = T \frac{d \ln \langle r^2 \rangle_0}{dT} \tag{6.16}$$

where $\langle r^2 \rangle_0$ represents the unperturbed dimensions of the network chains. This equation is of considerable importance for two reasons (Mark 1976). It permits the comparison of results of thermoelastic measurements on polymer chains in the bulk, in network structures, with results of viscosity measurements on chains of the same, essentially isolated, polymer in dilute solution. It also establishes the relationship between the purely thermodynamic quantity f_e/f and its molecular counterpart $d \ln \langle r^2 \rangle_0/dT$, which can be interpreted in terms of the rotational isomeric state theory of chain configurations (Flory 1969).

SOME EXPERIMENTAL DETAILS

It is, of course, important to note that use of the thermodynamic relationships derived in the previous section requires that the polymer network be brought as closely as possible to elastic equilibrium at each temperature of investigation. Thus only those thermoelastic data found to be reproducible upon subsequent changes in temperature are suitable for the purposes at hand. Swelling a network with nonvolatile solvent is frequently used to facilitate the approach to equilibrium.

Direct use of the equations pertaining to the condition of constant volume has the considerable advantage of not requiring recourse to an elastic equation of state of nonthermodynamic origin. Such experiments, however, generally require the imposition of a sizable hydrostatic pressure on the sample to keep its volume constant with changing temperature, and this requirement introduces considerable experimental difficulty (Allen et al. 1971). Because of this complication, such experiments have been restricted to very small temperature ranges, typically intervals of only a few degrees, thereby introducing additional uncertainties.

Measurements at constant pressure are much easier to carry out but require for their interpretation adoption of an elastic equation of state based on a particular model of the network, as already described. In the case of uniaxial elongation or compression at constant pressure, measurements at constant deformation α have an advantage in that calculation of values of f_e/f involves a correction term $\beta/3$, which is independent of deformation (Treloar 1975). The advantage of involving a deformation-independent correction term is shared by measurements in torsion or shear in general. (Measurements at constant α are of course somewhat inconvenient in that the temperature dependence of L_i requires readjustment of the length L of the sample at each new temperature, or the application of the appropriate correction for thermal expansion of the network in the unstrained state). Correction of measurements at constant length, however, requires the term $\beta/(\alpha^3 - 1)$, which tends to have very large values in the region of small deformation (Shen and Croucher 1975; Mark 1976). This term can be made negligibly small in elongation by choosing sufficiently large values of α, but no such simplification is possible in the case of compression (Mark 1976). Finally, thermoelastic data equivalent to those described above may, of course, be obtained from a series of stress–strain isotherms, or from measurements of sample length as a function of temperature at constant force (Mark 1976).

SOME TYPICAL DATA

Figure 9.1 shows some typical thermoelastic data, in elongation at constant length, directly represented as the dependence of force on temperature. From

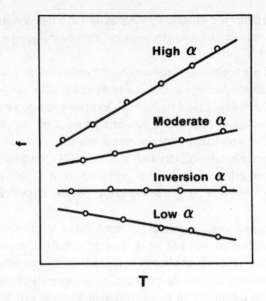

Figure 9.1 Typical thermoelastic results at various elongations.

the slopes of curves such as these, the derivatives required in Eq. (6.15), (9.2), and (9.3) may be obtained.

At low elongations, the increased tendency for the chains to retract (increase in f) with increase in temperature is overcome by the usual thermal expansion of the network (Treloar 1975). At a somewhat higher elongation, called the *inversion value*, the two effects just offset one another, and the force becomes independent of temperature. At still higher but moderate values of α, the usual easily measurable increase in f with T occurs, and this is the region where the most reliable thermoelasticity data are obtained. Very high elongations have the advantage of making the correction term $\beta/(\alpha^3 - 1)$ negligibly small, but they also have the possible disadvantage of introducing complications from strain-induced crystallization or from non-Gaussian effects due to limited chain extensibility (Mark 1979a,b).

SOME TYPICAL RESULTS

Table 9.1 summarizes some typical values of f_e/f and $d\ln\langle r^2\rangle_0/dT$ (Mark 1976). As can be seen, most values of these quantities are positive, which means that

Table 9.1   Some Typical Thermoelastic Results

Elastomer	$\dfrac{f_e}{f}$[a]	$10^3 \dfrac{d \ln \langle r^2 \rangle_0}{dT}$[b] (K^{-1})
Natural rubber	0.18	0.60
cis-1,4-Polybutadiene	0.13	0.44
Poly(dimethylsiloxane)	0.20	0.67
Polyisobutylene	−0.06	−0.20
Polyethylene	−0.42	−1.41
Elastin[c]	0.26	0.87

[a]Fraction of the force that is of energetic origin.
[b]Temperature coefficient of the unperturbed dimensions, in the vicinity of 298 K.
[c]An elastomeric protein present in mammals.
Source: Mark (1976).

extended chain conformations are generally of higher energy. Since an increase in temperature would then increase the frequency of these more extended conformations, this would also explain the positive values of $d \ln\langle r^2 \rangle_0/dT$. More detailed interpretations in terms of Rotational Isomeric State theory (Flory 1969) are discussed later in this chapter.

EVALUATION AND IMPORTANCE OF THERMOELASTIC RESULTS

Despite the experimental difficulties attending the study of unswollen polymer networks under the constraint of constant volume, a number of studies meeting this thermodynamic requirement have been successfully carried out by Allen and coworkers (Allen et al. 1971). Such studies are of great importance in evaluating the statistical theory of rubberlike elasticity, of course, since they permit the calculation of f_e/f directly from Eq. (6.15), a thermodynamically exact equation. Thermoelastic results thus obtained therefore serve as an important check of results obtained at constant pressure and interpreted by using the equation of state derived from the statistical theory.

Values of f_e/f obtained to date by the use of these methods are summarized in Table 9.2; included for purposes of comparison are the most reliable values of f_e/f obtained for the same polymer networks from measurements at constant pressure (Treloar 1975; Mark 1976). For the three polymers thus studied, val-

Table 9.2   Values of f_e/f Obtained at Constant Volume and Constant Pressure

Elastomer	Constant V^a	Constant p^a
Natural rubber	0.12	0.17
Poly(dimethylsiloxane)	0.25	0.20
Polyisobutylene	−0.09	−0.06

[a]Data from Treloar (1975); Mark (1976).

ues of f_e/f obtained from constant-pressure experiments and treated according to Eq. (9.2) or (9.3) are generally in good agreement with those obtained at constant volume.

Although the great majority of thermoelasticity studies have been carried out on networks in simple uniaxial elongation, several studies have now been reported in which the deformation was uniaxial compression, torsion, shear, or isotropic swelling. Results obtained for some polymers that have been studied in more than one type of deformation are presented in Table 9.3 (Treloar 1975; Mark 1976). The fact that results obtained from measurements in elongation are in good agreement with those from other types of deformation is of considerable importance since interchain ordering, if present, could be expected to respond differently to different types of network deformation and thus manifest itself through a dependence of f_e/f on the type of deformation employed. No such effect is evident from the results. Similarly, f_e/f has been found to be independent of cross-linking conditions, degree of cross-linking, and extent of deformation (Mark 1976).

Direct calorimetric measurements carried out on a polymer network during the deformation process can also be used to determine values of f_e/f. Some of

Table 9.3   Values of f_e/f Obtained Using Different Types of Deformation

Polymer	Elongation	Compression	Torsion	Shear[a]	Swelling Equilibrium
Natural rubber	0.17	0.18	0.17	0.18	—
cis-1,4-Polybutadiene	0.12	0.14	0.10	—	—
Poly(dimethylsiloxane)	0.20	0.13	—	—	~0.1
Polyisobutylene	−0.06	—	−0.03	—	—

[a]Elongation data converted into shear moduli.

Sources: Treloar (1975); Mark (1976).

Table 9.4 Comparison of Calorimetric and Thermoelastic Values of f_e/f

Elastomer	Type of Deformation	Calorimetric Values[a]	Thermoelastic Values[a]
Natural rubber	Elongation	0.18	0.17
	Torsion	0.20	0.15
cis-1,4-Polybutadiene	Elongation	0.11	0.12
	Torsion	0.14	0.11

[a]Data from Treloar (1975); Mark (1976).

the most reliable calorimetric values thus obtained are given in Table 9.4 (Treloar 1975; Mark 1976), and are seen to be in good agreement with thermoelastic results reported for the same two polymers.

One of the most direct confrontations of theory with experiment may be achieved by the thermoelastic study of networks swollen with diluent, either at constant composition or in swelling equilibrium. The effects of such diluents on f_e/f for several elastomers are summarized in Table 9.5 (Mark 1976). A wide variety of polymers and diluents have been employed in such studies, and a wide range of degrees of swelling have been investigated, down to a volume fraction of polymer of 0.18. Nonetheless, there is no apparent effect of dilution on f_e/f. This is an important result, since any ordering of polymer chains in the amorphous state would certainly be dispersed by the relatively large amounts of diluent incorporated into the networks in these studies (Flory 1973, Mark 1984a). It also supports the assumption that intermolecular interactions have little effect on rubberlike elasticity.

Conversion of the thermodynamic quantity f_e/f to $d \ln \langle r^2 \rangle_0 / dT = f_e / fT$ permits comparison of values of the temperature coefficient of the unperturbed

Table 9.5 Constancy of f_e/f upon Swelling

Elastomer	Unswollen Networks	Swollen Networks
Natural rubber	0.17	0.16
cis-1,4-Polybutadiene	0.12	0.11
trans-1,4-Polyisoprene	−0.10	−0.17
Polyethylene	−0.42	−0.52

Source: Mark (1976).

dimensions of the network chains as obtained from thermoelasticity measurements with values obtained from intrinsic viscosity–temperature ($[\eta] - T$) measurements on chains of the same polymer dispersed in a solvent at infinite dilution. Table 9.6 (Mark 1976) lists some polymers that have been studied both in thermoelastic investigations, and in viscometric studies in which proper account has been taken of polymer–solvent interactions and their variation with temperature. The excellent agreement obtained between values of $d \ln \langle r^2 \rangle_0 / dT$ for the same polymer chains under such vastly different conditions is extremely important since it is inconceivable that such agreement would be obtained if intermolecular interactions varied with configuration or if significant molecular ordering occurred in the amorphous state (Flory 1973; Mark 1984a).

Since intermolecular interactions do not affect the force, they must be independent of the extent of the deformation and thus the spatial configurations of the chains. This in turn indicates that the spatial configurations must be independent of intermolecular interactions; that is, the amorphous chains must be in random, unordered configurations, the dimensions of which should be the unperturbed values. This conclusion has now been amply verified, in particular by neutron scattering studies on undiluted amorphous polymers (Flory 1973).

The major conclusions reached in thermoelastic studies are summarized in Table 9.7.

ROTATIONAL ISOMERIC STATE INTERPRETATION

In the case of polyethylene, the thermoelastic data indicate that the energetic contribution to the elastic force is large and negative (Flory 1969). These results may be understood using the information given in Figure 9.2 (Mark 1981). The

Table 9.6   Comparison of Thermoelastic and Viscometric Values of Temperature Coefficient of Unperturbed Dimensions

Polymer	$10^3 d \ln \langle r^2 \rangle_0 / dT$	
	Thermoelastic	Viscometric
Poly(dimethylsiloxane)	0.59	0.52
Polyisobutylene	−0.19	−0.26
Polyethylene	−1.05	−1.06
Isotactic poly(n-pentene-1)	0.34	0.52

Table 9.7 Summary and Conclusions, Thermoelasticity Studies

A. Evidence
   1. Values of $f_e/f = T\, d \ln \langle r^2 \rangle_0/dT$ are independent of cross-linking conditions, degree of cross-linking, type and extent of deformation, and presence of diluent in a network.
   2. Values of $d \ln \langle r^2 \rangle_0/dT$ from thermoelastic studies are in good agreement with values from $[\eta]-T$ experiments.

B. Conclusions
   1. Intermolecular interactions must be independent of the extent of deformation (i.e., f_e/f must be *intramolecular*).
   2. Chains in the bulk amorphous state must be in random, *unordered* configurations.
   3. The molecular representation of the results is well interpreted by rotational isomeric state theory.

preferred (lowest-energy) conformation of the chains is the all-*trans* form, since gauche states (at rotational angles of $\pm 120°$) cause steric repulsions between CH_2 groups. Since this conformation has the highest possible spatial extension, stretching a polyethylene chain requires switching some of the gauche states (which are of course present in the randomly coiled form) to the alternative trans states. These changes decrease the conformational energy and are thus the origin of the negative type of ideality represented in the experimental value of f_e/f. [This physical picture also explains the decrease in unperturbed dimensions

A. RESULTS

$$\frac{f_e}{f} = T\, \frac{d \ln \langle r^2 \rangle_0}{dT} = -0.42$$

B. INTERPRETATION

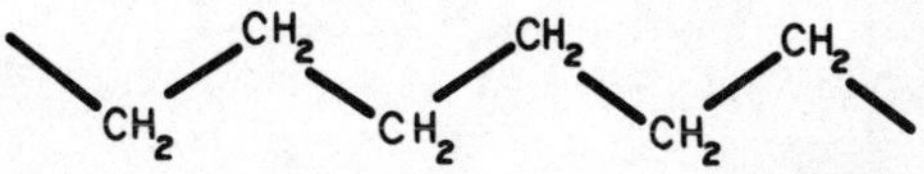

Figure 9.2 Thermoelastic results on (amorphous) polyethylene networks and their interpretation in terms of the preferred all-trans conformation of the chain (Mark 1981).

A. RESULTS

$$\frac{f_e}{f} = T \frac{d \ln \langle r^2 \rangle}{dT} = 0.20$$

B. INTERPRETATION

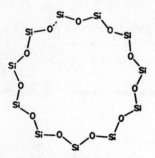

Figure 9.3 Thermoelastic results on poly(dimethylsiloxane) networks and their interpretation in terms of the preferred all-trans conformation of the chain (Mark 1981). For purposes of clarity, the two methyl groups on each silicon atom have been deleted.

upon increase in temperature. The additional thermal energy causes an increase in the number of the higher-energy gauche states, which are more compact than the trans ones (Flory 1969).]

The opposite behavior is observed in the case of poly(dimethylsiloxane) (PDMS), as is described in Figure 9.3 (Mark 1981). The all-trans form is again the preferred conformation; the relatively long Si–O bonds and the unusually large Si–O–Si bond angles reduce steric repulsions in general, and the trans conformation places CH_3 side groups at distances of separation where they are strongly attractive. Because of the inequality of the Si–O–Si and O–Si–O bond angles, however, this conformation is of very low spatial extension. Stretching a poly(dimethylsiloxane) chain therefore requires an increase in the number of gauche states. Since these are of higher energy, this explains the fact that deviations from ideality for these networks are found to be positive (Flory 1969).

As here illustrated, rotational isomeric state theory can be used to give an insightful molecular interpretation of the thermoelastic properties of polymer networks.

10

FORCE AS A FUNCTION OF STRUCTURE

INTRODUCTION

Until recently, there was relatively little reliable quantitative information on the relationship between stress and network structure, primarily because of the uncontrolled manner in which elastomeric networks were generally prepared (Flory 1953; Coran 1978). Specifically, segments close together in space were linked irrespective of their locations along the chain trajectories, thus resulting in a highly random network structure in which the number and locations of the cross-links were essentially unknown. Such a structure is shown in Figure 1.2.

New synthetic techniques are now available, however, for the preparation of "model" polymer networks of known structure (Mark 1981, 1985a; Gottlieb et al. 1981; Queslel and Mark 1984; Miller and Macosko 1987). An example is the reaction shown in Figure 10.1, in which hydroxyl-terminated PDMS chains are end-linked using tetraethyl orthosilicate. Characterizing the uncross-linked chains with respect to molecular weight M_n and molecular weight distribution and then running the specified reaction to completion gives elastomers in which the network chains have these characteristics. In particular they have a molecular weight M_c between cross-links equal to M_n, and a distribution the same as that of the precursor chains. The cross-links, of course, have the functionality of the end-linking agent (Mark and Sullivan 1977).

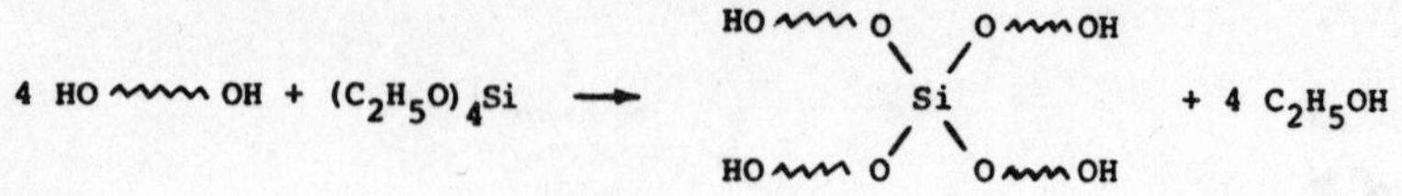

4 HO 〰〰 OH + (C$_2$H$_5$O)$_4$Si ⟶

$$\begin{array}{c} \text{HO} \sim\!\!\sim\!\!\sim\text{O} \quad \text{O} \sim\!\!\sim\!\!\sim\text{OH} \\ \diagdown \; \diagup \\ \text{Si} \\ \diagup \; \diagdown \\ \text{HO} \sim\!\!\sim\!\!\sim\text{O} \quad \text{O} \sim\!\!\sim\!\!\sim\text{OH} \end{array}$$

+ 4 C$_2$H$_5$OH

where HO〰〰OH represents a hydroxyl-terminated PDMS chain.

Known $\overline{M}_n$ ⟶ known $\overline{M}_c$.   Known M_n distribution ⟶ known M_c distribution.

Figure 10.1   A typical synthetic route for preparing elastomeric networks of known structure. (Reprinted with permission from J. E. Mark et al., Eds., *Physical Properties of Polymers*. Copyright 1984 American Chemical Society.)

EFFECTS OF CROSS-LINK FUNCTIONALITY

Trifunctional and tetrafunctional PDMS networks prepared in this way have been used to test the molecular theories of rubber elasticity with regard to the increase in nonaffineness of the network deformation with increasing elongation. Some of these results are shown in Figure 10.2. As can be seen, the ratio $2C_2/2C_1$ described in Chapter 8 decreases with increase in cross-link functionality from 3 to 4.

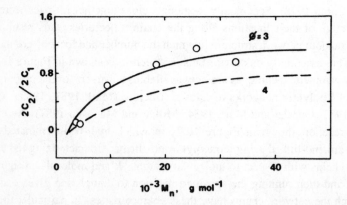

Figure 10.2   Experimental data showing values of the ratio $2C_2/2C_1$, which is a measure of the increase in nonaffineness of the deformation as the elongation increases (Mark et al. 1979). The ratio decreases with increase in junction functionality and with decrease in network chain molecular weight M_n, as predicted by theory.

A more thorough investigation of the effects of cross-link functionality required use of the more versatile chemical reaction illustrated in Figure 10.3. Specifically, vinyl-terminated PDMS chains were end-linked using a multifunctional silane. In the study summarized in Figure 10.4, this reaction was used to prepare PDMS model networks having functionalities ϕ ranging from 3 to 11, with a relatively unsuccessful attempt to achieve a functionality of 37. As shown in the figure, the modulus $2C_1$ (which is an estimate of the phantom limit for $[f^*]$) increases with increase in functionality. This is to be expected from an increase in the factor $(1 - 2/\phi)$ mentioned in Chapter 6. Also, $2C_2$ and its value relative to $2C_1$ both decrease. This results from the fact that the more chains there are emanating from a cross-link, the more constrained it is. There is therefore less of a decrease in modulus brought about by the fluctuations that are enhanced at high deformation and give the deformation its nonaffine character (Llorente and Mark 1980).

EFFECTS OF MOLECULAR WEIGHT BETWEEN CROSS-LINKS

The decrease in $2C_2/2C_1$ with decrease in network chain molecular weight also illustrated by the two curves in Figure 10.2 is due to the fact that there is less configurational interpenetration in the case of short network chains (Mark et al. 1979). This was discussed in Chapter 5. The decrease in interpenetration decreases the firmness with which the cross-links are embedded, and thus the deformation is already highly nonaffine even at relatively small deformations.

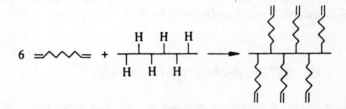

Figure 10.3 A typical reaction in which vinyl-terminated PDMS chains are end-linked with a multifunctional silane (Mark 1984a). (Reprinted with permission from J. E. Mark et al., Eds., *Physical Properties of Polymers*. Copyright, 1984 American Chemical Society.)

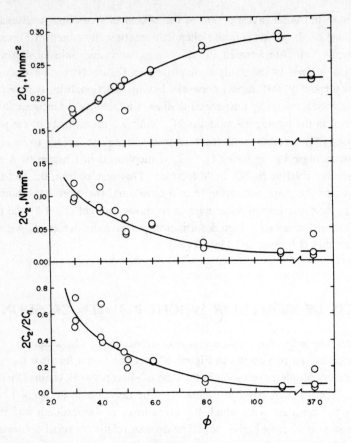

Figure 10.4 Experimental data showing the effect of cross-link functionality on $2C_1$ (a measure of the high-deformation modulus), $2C_2$, and $2C_2/2C_1$ (measures of the extent to which the nonaffineness of the deformation increases with elongation). (Reprinted with permission from M. A. Llorente and J. E. Mark, *Macromolecules*, **13**, 681. Copyright, 1980 American Chemical Society.)

EFFECTS OF INTERCHAIN ENTANGLEMENTS

Such model networks may also be used to provide a direct test of molecular predictions of the modulus of a network of known degree of cross-linking. Some experiments on model networks have given values of the elastic modulus in good agreement with theory. Others have given values significantly larger than

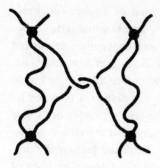

Figure 10.5 A "trapped" interchain entanglement. (Reprinted with permission from J. E. Mark et al., Eds., *Physical Properties of Polymers.* Copyright 1984 American Chemical Society.)

predicted, and the increases in modulus have been attributed to contributions from "permanent" chain entanglements (Langley 1968; Ziabicki 1976) of the type shown in Figure 10.5 and in the lower right corner of Figure 1.2. There are disagreements, and the issue has not yet been entirely resolved. Since the relationship of modulus to structure is of such fundamental importance, there is currently a great deal of research activity in this area (Mark 1981, 1985a; Gottlieb et al. 1981; Queslel and Mark 1984; Miller and Macosko 1987; Brotzman and Flory 1987).

In principle, solid-state nuclear magnetic resonance (NMR) spectroscopy could resolve this issue by in effect directly counting the cross-links. For example, NMR spectra of small siloxane molecules of the type $\wedge$ and $+$ have shown where the resonances should appear when the corresponding fragments appear as trifunctional and tetrafunctional cross-links, respectively. The absorbance at these frequencies then gives an estimate of the number density of cross-links and therefore leads to an independent prediction of the elastic modulus. Some success was achieved for very heavily cross-linked PDMS networks but not for the lower cross-link densities characteristic of elastomeric materials (Beshah et al. 1986). Also, similar experiments on alkane molecules were begun but, at least so far, not completed (Bennett et al. 1976).

Cross-linking in solution decreases the degree of entangling in the vicinity of the cross-links, as is described in Chapter 11.

EFFECTS OF DANGLING-CHAIN IRREGULARITIES

Since dangling chains represent imperfections in a network structure, one would expect their presence to have a detrimental effect on an elastomer's ultimate properties, specifically the elongation α_r and nominal or engineering stress $(f/A_d)_r$, at rupture. This expectation is confirmed by an extensive series of results obtained on PDMS networks that had been tetrafunctionally cross-linked using a variety of techniques (Andrady et al. 1981). Some pertinent results are shown, as a function of the molecular weight between cross-links, in Figure 10.6. The largest values of $(f/A_d)_r$ are obtained for the networks prepared by selectively joining functional groups occurring either as chain ends or as side groups along the chains. This is to be expected, because of the relatively low incidence of dangling ends in such networks. (The effects are particularly pronounced when such model networks are prepared from a bimodal mixture of relatively long and very short chains, as described in Chapter 13.) Also as expected, the lowest values of the ultimate properties generally occur for the networks cured by radiation (UV light, high-energy electrons, and γ radiation). The peroxide-cured networks are generally intermediate to these two extremes,

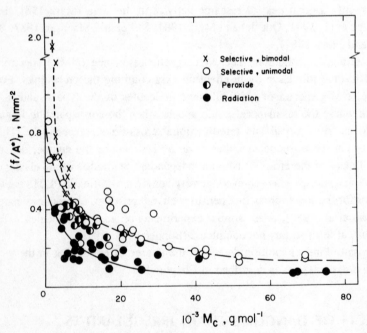

Figure 10.6 Values of the ultimate strength shown as a function of the molecular weight M_c between cross-links for (unfilled) tetrafunctional PDMS networks at 25°C (Andrady et al. 1981). The undeformed cross-sectional area A^* in this figure and in Figure 10.8 is the same as A_d in Eq. (6.13).

with the ultimate properties presumably depending on whether or not the free radicals generated by the peroxide are sufficiently reactive to cause some chain scission. Similar results were obtained for the maximum extensibility (Andrady et al. 1981). These observations are at least semiquantitative and certainly interesting, but they are somewhat deficient in that information on the number of dangling ends in these networks is generally not available.

More definitive results may be obtained by generating these irregularities in a more carefully controlled manner. Specifically, model networks containing dangling chains of known lengths and concentrations can be prepared in several ways, two of which are shown in Figure 10.7 (Mark 1985a). If, during the end-linking process, more difunctional chains are present than are required to react with all the functional groups on the end-linking molecules, then the known excess number of chain ends is equal to the number of dangling ends. In this method, the dangling chains must, of course, have the same average length as the elastically effective chains. The second method overcomes this limitation by including monofunctional chains of any desired length. In this way the dangling chains can be either much shorter or much longer than the elastically effective chains. A mixture of dangling chain lengths can also be introduced, as is, in fact, shown in Figure 10.7*b*.

An investigation using the first approach involved a series of model networks prepared by end-linking vinyl-terminated PDMS chains (Andrady et al. 1981). The tetrafunctional end-linking agent was used in varying amounts, all of which

(a)   (b)

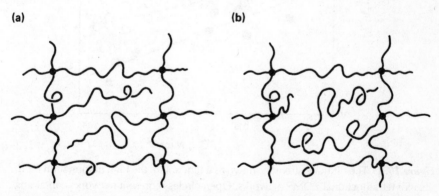

Figure 10.7 Two methods for preparing networks having dangling chains of known length, present in known concentrations. (a) Excess difunctional chains; (b) monofunctional chains. (Reprinted with permission from J. E. Mark, *Acc. Chem. Res.*, **18**, 202. Copyright 1985 American Chemical Society.)

were smaller than the amount corresponding to a stoichiometric balance between its active hydrogen atoms and the chains' terminal vinyl groups. The ultimate properties of these networks, with known numbers of dangling ends, were then compared with the properties obtained from networks previously prepared so as to have negligible numbers of these irregularities.

Values of the ultimate strength of the networks are shown as a function of the high deformation modulus $2C_1$ in Figure 10.8. The networks containing the dangling ends have lower values of $(f/A^*)_r$, with the largest differences occurring at high proportions of dangling ends (low $2C_1$ values), as expected. These results thus confirm the less definitive results shown in Figure 10.6. The values of the maximum extensibility show a similar dependence, as expected (Andrady et al. 1981).

EFFECTS OF INTERPENETRATING STRUCTURES

If two types of chains have different end groups, it is possible to end-link them simultaneously into two networks that interpenetrate one another (Sperling

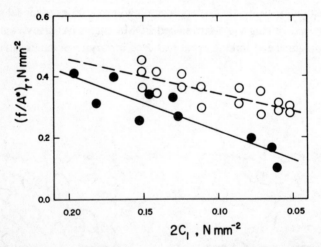

Figure 10.8 The ultimate strength shown as a function of the high deformation modulus for tetrafunctional PDMS networks. Open circles represent networks containing a negligible number of dangling ends; solid circles represent networks containing dangling ends introduced by using less than the stoichiometrically required amount of end-linking agent (Andrady et al. 1981). In the latter case, decrease in $2C_1$ corresponds to increase in the number of dangling ends.

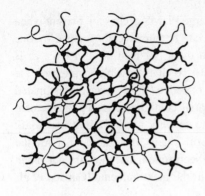

Figure 10.9 A bimodal interpenetrating network, in which the short-chain cross-links are represented by filled circles and the long-chain cross-links by open circles. (Reprinted with permission from J. E. Mark, *Acc. Chem. Res.*, **18**, 202. Copyright 1985 American Chemical Society.)

1981). Such a network could, for example, be made by reacting hydroxyl-terminated PDMS chains with tetraethylorthosilicate while reacting vinyl-terminated PDMS chains with a multifunctional silane (Mark and Ning 1985). A bimodal network of this type is shown in Figure 10.9. Interpenetrating networks in general can be very unusual with regard to both equilibrium and dynamic mechanical properties. For example, such materials can have considerable toughness and unusual damping characteristics.

EFFECTS OF LINEAR AND CYCLIC DILUENTS

End-linking functionally terminated chains in the presence of chains with inert ends yields networks through which the unattached chains *reptate* (de Gennes 1979). Networks of this type have been used to determine the efficiency with which unattached chains can be extracted from an elastomer as a function of their lengths and the degree of cross-linking of the network (Garrido and Mark 1985). The efficiency is found to decrease with increase in molecular weight of the diluent and with increase in degree of cross-linking, as expected. It has also been found to be more difficult to extract diluent present during the cross-linking than to extract the same diluents absorbed into the networks after cross-linking. Such comparisons can provide valuable information on the arrangements and transport of chains within complex network structures.

Some aspects of the stress–strain behavior of swollen networks that were cross-linked in the undiluted state are described in Chapter 8. Some properties of unswollen networks that were cross-linked in solution are given in Chapter 11.

It has also been found that if relatively large PDMS cyclics are present when linear PDMS chains are end-linked, some can be permanently trapped by one or more network chains threading through them (Clarson et al. 1986), as is shown by cyclics B, C, and D in Figure 10.10. It is possible to interpret the fraction of cyclics trapped in terms of their effective "hole" sizes, as estimated from Monte Carlo simulations of their spatial configurations. The agreement between theory and experiment is found to be very good (DeBolt and Mark 1987a). Since the cyclics constrain the network chains, they should increase the modulus of the elastomer. Increases have been found, but they are not much larger than the experimental uncertainties in the measurements (Garrido et al. 1985b).

It is also possible to use this technique to form a network having no cross-links whatsoever. Mixing linear chains with large amounts of cyclic and then *di*functionally end-linking them can give sufficient cyclic interlinking to yield an *Olympic* or *chain-mail* network (de Gennes 1979; Rigbi and Mark 1986), as

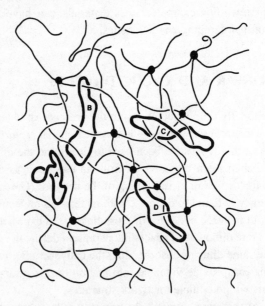

Figure 10.10 A tetrafunctional PDMS network containing cyclics (heavy lines). Cyclics B, C, and D were trapped by linear chains that passed through them prior to being end-linked into the network. (Reprinted with permission from L. C. DeBolt and J. E. Mark, *Macromolecules*, **20**, 2369. Copyright 1987 American Chemical Society.)

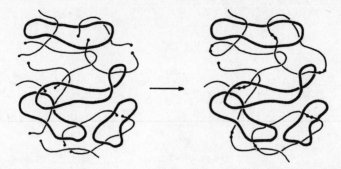

Figure 10.11 Preparation of a chain-mail network, which has no cross-links at all (Garrido et al 1985a). Linear chains (light lines) passing through the cyclics (heavy lines) are difunctionally end-linked to form a series of interpenetrating cyclics. The result is a gel.

is illustrated in Figure 10.11 (Garrido et al. 1985a). Such materials could have highly unusual stress–strain isotherms, and attempts to carry out the required characterization measurements are in progress.

PART B

ADDITIONAL TOPICS

11

NETWORKS PREPARED UNDER UNUSUAL CONDITIONS

INTRODUCTION

As can be gathered from Chapter 10, it is very difficult to obtain information on the topology of a network. Some studies have therefore taken an indirect approach. Networks were prepared in a way that could be expected to simplify their topologies, and their properties were measured and interpreted in these terms.

THE EXPERIMENTAL APPROACHES

Two techniques that may be used to prepare networks having simpler topologies are illustrated in Figure 11.1. Basically, they involve separating the chains prior to their cross-linking by either stretching (Greene et al. 1965) or dissolution (Johnson and Mark 1972). After the cross-linking, the stretching force or solvent is removed and the network is studied (unswollen) with regard to its stress–strain properties in elongation.

SOME RESULTS AND THEIR QUALITATIVE INTERPRETATION

Some results obtained on PDMS networks cross-linked in solution by means of γ radiation are shown in Table 11.1 (Johnson and Mark 1972; Mark 1984a).

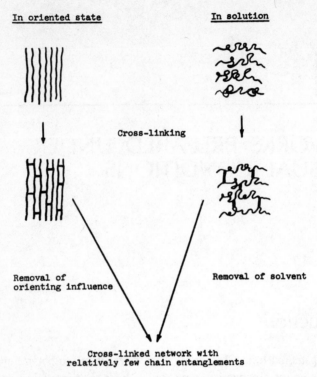

<u>In oriented state</u>   <u>In solution</u>

Cross-linking

Removal of
orienting influence

Removal of solvent

Cross-linked network with
relatively few chain entanglements

Figure 11.1 Two techniques that may be used to prepare networks of simpler topology. (Reprinted with permission from J. E. Mark et al., Eds., *Physical Properties of Polymers.* Copyright 1984 American Chemical Society.)

Note the continual decrease in the time required to reach elastic equilibrium, t_{eq}, and in the extent of stress relaxation as measured by the ratio of equilibrium to initial values of $[f^*]$, upon decrease in the volume fraction of polymer present during the cross-linking. Also, at higher dilutions there is a further decrease in the Mooney–Rivlin $2C_2$ constant described in Chapter 8.

These observations are qualitatively explained in Figures 11.2 and 11.3. If a network is cross-linked in solution and the solvent then removed, the chains collapse in such a way that there is reduced overlap in their configurational domains. As a result, as can be seen from Chapter 5, κ is decreased. It is primarily in this regard—namely, decreased chain–cross-link entangling—that solution-cross-linked samples have simpler topologies, with correspondingly simpler elastomeric behavior.

Table 11.1 PDMS Networks Compared at Approximately Constant Modulus

$v_{2,s}{}^a$	t_{eq} (hr)	$\dfrac{[f^*](equil)}{[f^*](init)}$	$2C_2$ (N mm^{-2})
1.00	0.70	0.95	0.062
0.75	0.48	0.98	0.057
0.62	0.10	0.99	0.059
0.55	0.30	0.99	0.062
0.48	0.02	1.00	0.067
0.40	0.03	1.00	0.039
0.30	0.00	1.00	0.031

^{a}Volume fraction of polymer in the solution being irradiated in the cross-linking reaction, which is generally somewhat larger than the volume fraction v_{2C} actually successfully incorporated in the gel.

Source: Johnson and Mark (1972); Mark (1984a). Reprinted with permission from J. E. Mark et al., Eds., *Physical Properties of Polymers.* Copyright 1984 American Chemical Society.

Figure 11.2 Typical configurations of four chains emanating from a tetrafunctional cross-link in a polymer network prepared in the undiluted state. (Reprinted with permission from J. E. Mark et al., Eds., *Physical Properties of Polymers.* Copyright 1984 American Chemical Society.)

Figure 11.3 Typical configurations of four chains emanating from a tetrafunctional cross-link in a (dried) polymer network that had been prepared in solution. (Reprinted with permission from J. E. Mark et al., Eds., *Physical Properties of Polymers*. Copyright 1984 American Chemical Society.)

INTERPRETATION IN TERMS OF THE CONSTRAINED JUNCTION THEORY

The theory is found to give a good account of these results when the constraint parameter κ is varied with volume fraction v_{2C} of polymer present during the cross-linking (Erman and Mark 1987). The values of κ obtained are within the range of values obtained in other comparisons of theory and experiment, as are the values of an additional, relatively unimportant heterogeneity parameter ζ (Flory and Erman 1982). The values of κ generally decrease with decrease in v_{2C} and with increase in degree of cross-linking as represented by the constant $2C_1$ described in Chapter 8. The dependence of κ on v_{2C} is significantly stronger than that suggested by theory, however, indicating a particularly strong effect of dilution on the degree of network chain interpenetration.

12

STRAIN-INDUCED CRYSTALLIZATION AND ULTIMATE PROPERTIES

SOME GENERAL COMMENTS ON CRYSTALLIZATION

As already mentioned in Chapter 1, large amounts of crystallization in an undeformed network interfere with rubberlike behavior because they suppress the mobility of the network chains. They can also enhance the degree of intermolecular correlations. A relatively small number of very small crystallites present in an undeformed network are no problem, however, and can in fact function as temporary cross-links, as described in Chapter 3.

Even better are polymers with melting points a little below room temperature, since these materials can undergo strain-induced crystallization at typical temperatures of utilization. As mentioned in Chapter 1, this type of crystallization occurs because the melting point $T_m = \Delta H_m / \Delta S_m$ is elevated by the decrease in entropy of the stretched network chains. Specifically,

$$\Delta S_m = S_{\text{amorph}} - S_{\text{cryst}} \tag{12.1}$$

It is S_{amorph} that is decreased by the stretching, and, since $S_{\text{cryst}} = 0$ (at least for perfect order), ΔS_m is decreased as well, and T_m increases.

The elevation of the melting point of natural rubber with stretching is shown schematically in Figure 12.1 (Mark 1984b). The melting point of very carefully annealed natural rubber is as high as 28°C, but under most conditions the melt-

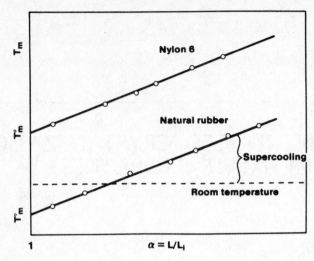

Figure 12.1 Comparison of strain-induced elevation of the melting point T_m° of an elastomer (natural rubber) and of a fiber (nylon 6). A typical extent of supercooling relative to room temperature is indicated for the elastomer. (Reprinted with permission from J. E. Mark and G. Odian, Eds., *Polymer Chemistry*. Copyright 1984 American Chemical Society.)

ing point T_m° in the unstretched state is below room temperature. Thus unstretched natural rubber generally is totally amorphous at room temperature. Stretching a network of the polymer, however, increases T_m considerably. The extent to which it exceeds the test temperature (for example, room temperature) is called *supercooling*, as is illustrated in the figure. Crystallization now occurs (is "induced"), as is readily shown by X-ray diffraction measurements. Permitting the natural rubber to retract decreases T_m back to T_m°, and the crystallites melt since the melting point is again back below room temperature.

Rather different behavior is observed for a polymer (such as nylon 6) having a melting point T_m° above room temperature. Stretching can now induce additional crystallization, but this crystallinity will persist even upon retraction of the polymer, as is also shown in the figure.

It should be mentioned that a type of *law of corresponding states* applies to this phenomenon. For example, if room temperature were somewhat lower (as, for example, on another planet), natural rubber would be very similar to nylon 6 in that they would both be fibrous at this new ambient temperature. Similarly,

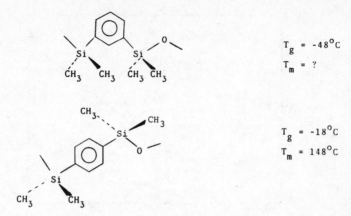

Figure 12.2 Structures, glass transition temperatures, and melting points of two silphenylene polymers. (a) Meta, and (b) para structures.

if room temperature were much higher, then nylon 6 would be very similar to natural rubber, in that under these conditions they would both be elastomers.

Strain-induced crystallization can reinforce an elastomer, as is described below. It is an advantage that natural rubber enjoys over silicone elastomers, which generally cannot exhibit strain-induced crystallization because of their very low melting points ($\sim -40°C$) in the undeformed state. There are now attempts to increase $T_m^°$ of silicone-type elastomers by stiffening the chains. This is illustrated by the two silphenylene polymers shown in Figure 12.2. The para phenylene group (Fig. 12.2b) is seen to have the larger stiffening effect, as expected: it causes the larger increase in glass transition temperature and increases the melting point to well above room temperature. No melting point has yet been reported for the meta polymer, but it could well undergo strain-induced crystallization. From the elastomeric point of view, this would make it much more important than the para polymer, which is obviously a thermoplastic at room temperature.

UPTURNS IN THE REDUCED STRESS AT HIGH ELONGATIONS

As already described in Figure 1.7, some (unfilled) networks show a large and rather abrupt increase in modulus at high elongations. This increase is further

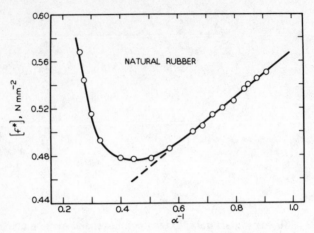

Figure 12.3 Stress–strain isotherm for an unfilled rubber network in the vicinity of room temperature showing the anomalous increase in modulus at high elongation (Mark et al., 1976; Mark 1979a).

illustrated for natural rubber in the Mooney–Rivlin representation in Figure 12.3. This behavior is very important in a practical sense since it corresponds to a significant toughening of the elastomer. Its molecular origin, however, has been the source of considerable controversy. It has been widely attributed to the *limited extensibility* of the network chains, that is, to an inadequacy in the Gaussian distribution function (Treloar 1975). This potential inadequacy is readily evident in the exponential in Eq. 4.1, specifically that this function does not assign a zero probability to a configuration unless its end-to-end separation r is infinite. This limited-extensibility explanation was viewed with skepticism by some workers since the increase in modulus was generally observed only in networks that could undergo strain-induced crystallization (Mark 1979a,b). Such crystallization in itself could account for the increase in modulus, primarily because the crystallites thus formed would act as additional cross-links in the network.

Attempts to clarify the problem by using noncrystallizable networks were not convincing since such networks were incapable of the large deformations required to distinguish between the two possible interpretations. The issue has now been resolved, however, by the use of end-linked, noncrystallizable model PDMS networks, as described in Chapter 10. These networks have high extensibilities, presumably because of their very low incidence of dangling-chain network irregularities (Mark 1982a). They have particularly high extensibilities when they are prepared from a mixture of very short chains (around a few hundred g mol^{-1}) with relatively long chains (around 20,000 g mol^{-1}) (Mark 1985b,c). Apparently the very short chains are important because of their limited extensibilities, and the relatively long chains are important because of their ability to retard the rupture process. As described in Chapter 13, these bimodal networks show upturns that are diminished by neither increase in temperature nor swelling. This non-Gaussian effect is therefore due to limited chain extensibility. Such results will be extremely useful for the reliable evaluation of the various non-Gaussian theories of rubberlike elasticity.

Crystallizable networks such as natural rubber show upturns that do diminish with increase in temperature or swelling. These upturns are obviously caused by strain-induced crystallization (Mark 1979a,b). The crystallization is due to the elevation of the melting point by the decrease in the entropy of fusion, as described in this chapter and in Chapter 1. The increase in the melting point can be very substantial, with the elevated T_m frequently lying above the decomposition temperature of the polymer. The suppression of the upturn by the increase in temperature or presence of solvent is due to suppression of the crystallization. In the case of the increase in temperature, the effect is simply the melting of the crystallites. The solvent, on the other hand, has the indirect effect of lowering the melting point, as illustrated in Figure 12.4 for the simpler case of an uncross-linked polymer (Flory 1953; Mark 1984b). When the polymer chains leave the crystalline lattice, they can further disorder themselves by mixing with the solvent. This increases ΔS of the process and thus decreases T_m.

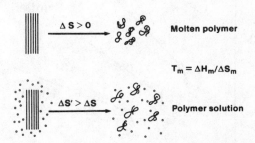

Figure 12.4 Sketch showing how the presence of a solvent (small open circles) can reduce the melting point of a polymer. (Reprinted with permission from J. E. Mark and G. Odian, Eds., *Polymer Chemistry*. Copyright 1984 American Chemical Society.)

DOWNTURNS IN THE REDUCED STRESS AT HIGH ELONGATIONS

In the case of crystallizable networks, there is frequently a downturn in the reduced stress just prior to its upturn, as illustrated in Figure 12.5 (Mark 1979a,b). The decrease in the upturns with swelling are due to suppression of some of the strain-induced crystallization by the solvent, as already described. Also, the initiation of the strain-induced crystallization (as evidenced by departure of the isotherm from linearity) is facilitated by the presence of the low-molecular-weight diluent, presumably by increasing the mobility of the network chains. Thus, in a sense this kinetic effect acts in opposition to the thermodynamic effect, which is primarily the suppression of the polymer melting point by the diluent. The most interesting point has to do with the decrease in the modulus prior to its increase. As shown schematically in Figure 12.6 (Mark 1979c), this is probably due to the fact that the crystallites are oriented along the direction of stretching (Fig. 12.6a), and the chain sequences within a crystallite are in regular, highly extended conformations (Fig. 12.6b). The straightening and aligning of portions of the network chains thus decrease the deformation in the remaining amorphous regions, with an accompanying decrease in the stress.

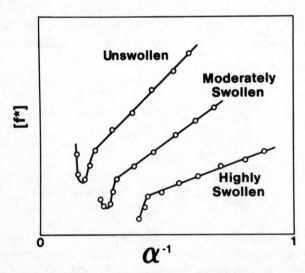

Figure 12.5 Stress–strain isotherms for a crystallizable network showing downturns in the reduced stress.

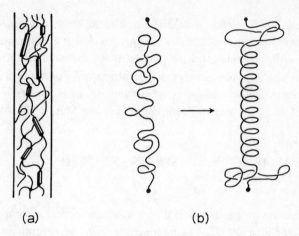

(a)   (b)

Figure 12.6 Strain-induced crystallization in a polymer network that has been elongated by a force along the vertical direction. Part (a) shows strain-induced crystallites, oriented preferentially along the fiber direction; (b) shows the straightening out of a portion of one of the network chains because of the crystallization. (Reprinted with permission from J. E. Mark, *Acc. Chem. Res.*, **12**, 49. Copyright 1979 American Chemical Society.)

ULTIMATE PROPERTIES

In this section we continue the discussion of unfilled elastomers at high elongations but emphasize ultimate properties, namely, the ultimate strength and maximum extensibility.

Some illustrative results on the effects of strain-induced crystallization on ultimate properties are given for *cis*-1,4-polybutadiene networks in Table 12.1

Table 12.1 Ultimate Properties for Crystallizable *cis*-1,4-Polybutadiene Networks

| | | Ultimate Properties | |
| | α at | Maximum Upturn | α at |
T (°C)	Upturn	in $[f^*]$ (%)	Rupture
5	3.27	54.2	6.64
10	3.48	30.1	6.22
25	4.03	4.3	5.85
40	—	0.0	5.68

Source: Su and Mark (1977); Mark (1984a).

(Su and Mark 1977; Mark 1984a). The higher the temperature, the lower the extent of crystallization and, correspondingly, the lower the ultimate properties. The effects of increase in swelling parallel those for increase in temperature, since diluent also suppresses network crystallization. For noncrystallizing networks, however, neither change is very important, as illustrated by the results shown for PDMS networks in Table 12.2 (Chiu and Mark 1977; Mark 1984a).

STATISTICAL THEORY OF STRAIN-INDUCED CRYSTALLIZATION

The basic features of the statistical theory of strain-induced crystallization have been given by Flory (1947). The treatment is based on several simplifying assumptions and is therefore of a qualitative nature, in contrast to the refined theories of the amorphous state. The assumptions and approximations of Flory's formulation are:

1. Crystallization is assumed to occur in a state of equilibrium. Such a state is reached experimentally, for example, by stretching the network at high temperature to a fixed length and subsequently decreasing the temperature to obtain crystallization. Most experiments on strain-induced crystallization are carried out at nonequilibrium conditions, where crystallization occurs while stretching.
2. Entropy changes associated with the formation of crystal nuclei are assumed to be negligible.
3. Crystallites are assumed to be oriented parallel to the axis of elongation.

Table 12.2   Ultimate Properties for Noncrystallizing PDMS Networks

v_2	$\lambda_r{}^a$	$[f^*]_r$
1.00	4.90	0.0362
0.80	4.42	0.0342
0.60	4.12	0.0338
0.40	4.16	0.0336

$^a\lambda_r$ is the value of the total elongation at the rupture point, using L_i (unswollen).

Source: Chiu and Mark (1977); Mark (1984a).

This assumption has subsequently been relaxed in a more refined treatment (Allegra 1980) in which crystallites are taken to form parallel to the end-to-end vector of each chain.

4. Chain folding is disregarded and each chain is assumed to traverse the crystallite only once, in the sense of direction of increasing chain vector. This assumption was removed later (Smith 1976) by permitting a chain to enter a crystallite in either direction. Chain folding has also been considered explicitly by Gaylord (1976).

5. Chains are assumed to be Gaussian.

The theory developed under these assumptions (Flory 1947) is applicable only under conditions of incipient crystallization. It gives explicit expressions for the degree of crystallinity and the incipient crystallization temperature as functions of the extension ratio, and for the applied force as a function of the extension ratio and the degree of crystallization. These relations are:

1. The fractional degree of crystallization ψ:

$$\psi = \left[\frac{\frac{3}{2} - \phi(\alpha)}{\frac{3}{2} - \theta} \right]^{1/2} \tag{12.2}$$

where α is the extension ratio in simple tension, and

$$\phi(\alpha) = \frac{2\alpha\beta l}{\pi^{1/2}} - \frac{\alpha^2/2 + 1/\alpha}{n} \tag{12.3}$$

$$\theta = (h_f/R)(1/T_m^\circ - 1/T) \tag{12.4}$$

Here, n is the number of segments in a freely jointed chain, l is the segment length, and $\beta = (3/2n)^{1/2}l$. The quantity h_f is the heat of fusion per segment, and T_m° is the incipient crystallization temperature for the undeformed polymer.

2. The incipient crystallization temperature T_m of the stretched network:

$$1/T_m = 1/T_m^\circ - (R/h_f)\phi(\alpha) \tag{12.5}$$

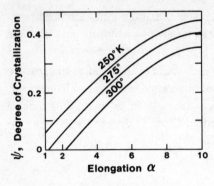

Figure 12.7 Degree of crystallization ψ at equilibrium as a function of elongation at the three temperatures indicated (Flory 1947).

3. The applied force, f^*:

$$f^* = \frac{\nu k T}{V_0} \frac{(\alpha - \alpha^{-2}) - (6n/\pi)^{3/2}\psi}{1 - \psi} \tag{12.6}$$

The relationship of the degree of crystallinity to the extension ratio given by Eq. (12.2) is shown in Figure 12.7 for three temperatures, and for $n = 50$, $h_f = 600R$, and $T_m^\circ = 250$ K. The failure of the $T = T_m^\circ = 250$ K curve to reach

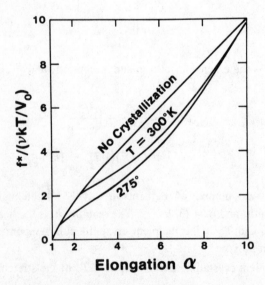

Figure 12.8 Stress–strain curves without crystallization and with equilibrium crystallization at the two temperatures indicated (Flory 1947).

zero crystallinity at $\alpha = 1$ results from the simplifying assumptions on which the theory is based.

The relationship of the dimensionless ratio $f^*/(\nu k T/V_0)$ to α obtained from Eq. (12.6) is depicted in Figure 12.8. The parameters used are the same as those for Figure 12.7. The decrease in force under equilibrium crystallization predicted by Eq. (12.6) and shown in Figure 12.8 may be identified with the downturns exhibited by the experimental data of Figure 12.5.

13

BIMODAL NETWORKS AND NON-GAUSSIAN BEHAVIOR

INTRODUCTION

As mentioned in Chapter 10, the use of very specific chemical reactions for the cross-linking process permits control of the network structure. The earlier discussion focused on controlling the average chain length in the preparation of model networks to test the molecular theories of rubberlike elasticity. Here, interest centers on preparing elastomers of controlled network chain length *distribution*. Of specific interest are end-linked bimodal networks consisting of very short chains and relatively long chains, the latter being representative of the usual unimodal networks. Figure 13.1 shows part of such a network.

Bimodal PDMS networks prepared in this way were first used to test the *weakest-link theory*, in which rupture was thought to be initiated by the shortest chains (because of their very limited extensibility) (Bueche 1962). In addition, in the unfilled state near room temperature, such networks were found to be unusually tough elastomers (Mark 1982a,b, 1985a). Of particular interest is the fact that they have values of [f^*] that increase very substantially at high elongations, thus giving unusually large values of the ultimate strength. This is rather surprising since usually an elastomer will have good ultimate properties only when reinforced with some mineral filler (or with hard, glassy domains in the case of a multiphase polymer), or when it can generate its own reinforcement through strain-induced crystallization (Smith 1977; Mark 1979a,b). Part

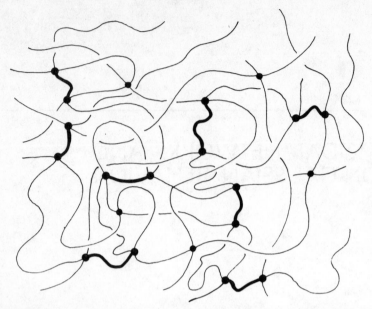

Figure 13.1 Sketch of a portion of a bimodal network (Mark 1979d). The very short polymer chains are arbitrarily represented by heavy lines and the relatively long chains by thinner lines. The dots represent cross-links.

of these increases in modulus and ultimate strength in bimodal networks are due to the low incidence of dangling-chain irregularities in such model networks in general. Another contribution could be non-Gaussian effects arising from limited chain extensibility, as discussed in Chapter 12. It is not known, however, whether these are the only effects occurring at very low temperatures. Such networks could also, for example, undergo strain-induced crystallization, with consequent improvements in ultimate properties from crystallite reinforcement. PDMS has a very low melting point ($-40°C$) and, although elongation significantly raises it (as described in Chapter 12), unfilled networks of this polymer are thought to remain amorphous well below room temperature. Unusually low temperatures must therefore be employed in any search for strain-induced crystallization. There are very few relevant data in the literature, and, of course, no relevant results whatever are available for the new types of networks with the very peculiar distributions of interest here. This question of possible strain-induced crystallization and the associated network reinforcement in the vicinity of room temperature assumes particular importance because of the very unusual and attractive properties of these materials.

A definitive answer to these questions can be obtained by analysis of how the relevant elastomeric properties depend on temperature, swelling, composition, chain length, spatial heterogeneity, and junction functionality. The resulting molecular interpretation of these unusual properties of bimodal networks permits the utilization of these elastomers in a variety of applications, a number of which are discussed in some detail (Mark 1982a, 1985a,b,c).

TEST OF THE WEAKEST-LINK THEORY

It was observed that increasing the number of very short chains in a bimodal network did not significantly decrease its ultimate properties. The reason, given schematically in Figure 13.2, is the very nonaffine nature of the deformation at such high elongations (Andrady et al. 1980). The network simply reapportions the increasing strain among the polymer chains until no further reapportioning is possible. It is generally only at this point that chain scission occurs, leading to rupture of the elastomer. The weakest-link theory (Bueche 1962) is implicitly based on the assumption of an affine deformation, which means that the elon-

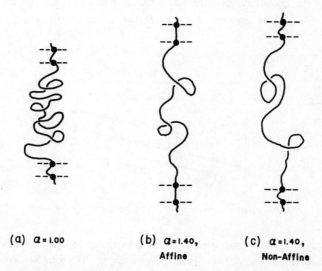

    (a) $\alpha = 1.00$ (b) $\alpha = 1.40$,   (c) $\alpha = 1.40$,
                                  Affine   Non-Affine

Figure 13.2 The effect of deformation on an idealized network segment consisting of a relatively long chain bracketed by two very short chains. (Reprinted with permission from A. L. Andrady et al., *J. Chem. Phys.*, **72**, 2282. Copyright 1980 American Chemical Society.)

gation at which the modulus increases should be independent of the number of short chains in the network. The experimental results show the opposite behavior; the smaller the number of short chains, the easier the reapportioning and the higher the elongation required to bring about the upturn in modulus (Andrady et al. 1980).

INHERENT TOUGHENING EFFECTS

For purposes of comparison, some typical stress–strain isotherms for a unimodal crystallizable elastomer are shown in Figure 13.3. Curve A, for a relatively low temperature, shows the upturn in $[f^*]$ characteristic of strain-induced crystallization (Mark 1979a,b). Increase in temperature (curve B) or addition of a swelling diluent (curve C) diminish or suppress this crystallization, thus removing the upturn in $[f^*]$, as described in Chapter 12. These curves are thus very similar to curve D, which is representative of the isotherms shown by unimodal elastomers that are inherently noncrystallizable.

Figure 13.4 shows some analogous results on bimodal networks, such as those of PDMS. Curve A, for a relatively low temperature shows an upturn as before, but the upturn is not diminished by an increase in temperature (Zhang and Mark 1982). As illustrated by Curve B, the values of $[f^*]$ simply increase,

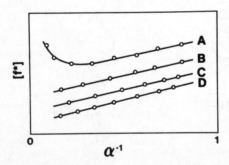

Figure 13.3 Schematic stress–strain isotherms in elongation for a unimodal crystallizable elastomer, in the Mooney–Rivlin representation. The top three are for a crystallizable network: curve A for a relatively low temperature, B for an increased temperature, and C for the introduction of a swelling diluent. Isotherm D is for an unswollen unimodal network that is also inherently noncrystallizable.

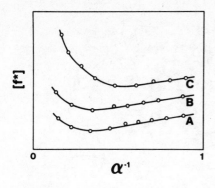

Figure 13.4 Schematic isotherms for a noncrystallizable bimodal network: curve A for a relatively low temperature, B for an increased temperature, and C for the introduction of a swelling diluent.

due to the usual increase $[f^*]$ shows with increasing temperature (as described in Chapters 4, 5, and 9). This suggests that the upturn in $[f^*]$ in this case is instead due to the limited extensibility of the very short chains, as does the effect of swelling. As can be seen from curve C, swelling can both increase the upturn and decrease the elongation required to bring it about (Mark 1985b). Additional experimental results on such bimodal elastomers are summarized below (Mark 1982a,b).

Experiments focusing on the effects of network composition showed that increase in the number of short chains in the networks generally gives a more pronounced increase in $[f^*]$ (Mark 1985b), thus underscoring the importance of the short-chain component in this regard. Properties improve up to a short-chain concentration of approximately 95 mol%. Beyond this point, properties decline because of increasing brittleness (Mark 1985c). If it is the short chains that give the improvements in ultimate strength, then decreasing their average length should give even more pronounced upturns in $[f^*]$ at high elongations. This is indeed found to be the case, and molecular weights as low as 220 g mol^{-1} have now been investigated.

Some insight into the mechanism through which the short chains operate may be obtained using bimodal networks that are made *spatially* as well as compositionally heterogeneous. This is effected in a two-stage reaction in which some of the very short chains are prereacted so as to form clusters or domains of high cross-link density (Mark and Andrady 1981). Such a network is shown schematically in Figure 13.5. If the increases in $[f^*]$ were due to some intermolecular organization such as a filler effect or strain-induced crystallization, then segregating the short chains should probably enhance the modulus. The ob-

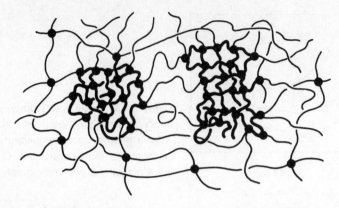

Figure 13.5   Sketch of a portion of a network that is spatially as well as compositionally heterogeneous with regard to chain length. Most of the two dense clusters of very short chains (heavy lines) were formed before the rest of the network (thin lines). (Mark and Andrady 1981.)

served increases are very small and in fact are within the usual error limits in these types of measurements. Thus there does not appear to be any significant reinforcing effect brought about by clustering the very short chains in the spatially inhomogeneous cross-linking process. This supports the idea that the increases in $[f^*]$ are primarily intramolecular rather than intermolecular. Also, such networks could serve as useful models for real networks inhomogeneously cross-linked, for example, by incomplete miscibility of a cross-linking agent.

The magnitude of the increase in $[f^*]$ does not show any obvious correlation with functionality ϕ, which again suggests the predominant importance of the intramolecular characteristics of the short chains. At least from the evidence at hand, the functionality of the junction points seems relatively unimportant in this regard.

With regard to additional possible effects of temperature, of particular interest are the temperature dependence of the elongation α_u at which the upturn starts, and of the total increase $\Delta[f^*]_r$ in modulus up to the rupture point. If the increase in $[f^*]$ had been due to strain-induced crystallization, α_u would have decreased with decrease in temperature, and α_r and $\Delta[f^*]_r$ would have increased. These qualities were found to be relatively insensitive to temperature (Zhang and Mark 1982).

Also relevant here are some force–temperature results obtained at elongations sufficiently large to give large increases in $[f^*]$ in the stress–strain isotherm. Such curves show no deviations from linearity that could be attributed to strain-induced crystallization. Similarly, birefringence–temperature measurements also carried out at $\alpha > \alpha_u$ show no deviations from linearity that could be attributed to crystallization, or to other intermolecular orderings of the network chains (Zhang and Mark 1982).

CHARACTERIZATION OF LIMITED CHAIN EXTENSIBILITY

Now that we have elucidated the molecular origin of the unusual properties of bimodal networks, at least to some extent, it is possible to interpret the limited chain extensibility in terms of the configurational characteristics of the PDMS chains making up the network structure.

The first important characteristic of limited chain extensibility is the elongation α_u at which the increase in $[f^*]$ first becomes discernible. Values of this minimum elongation are readily obtainable from the stress–strain isotherms. Although the deformation is nonaffine in the vicinity of the upturn, it is possible to provide at least a semiquantitative interpretation of such results in terms of the network chain dimensions (Andrady et al. 1980). At the beginning of the upturn, the average extension r of a network chain having its end-to-end vector along the direction of stretching is simply the product of the unperturbed dimension $\langle r^2 \rangle_0^{1/2}$ and α_u. Similarly, the maximum extensibility r_m is the product of the number n of skeletal bonds and the factor 1.34 Å that gives the axial component of a skeletal bond in the most extended helical form of PDMS, as obtained from geometric analysis. The ratio r/r_m at α_u thus represents the fraction of the maximum extensibility occurring at this point in the deformation. The values obtained indicate that the upturn in modulus generally begins at approximately 60–70% of maximum chain extensibility. This relatively reliable estimate is approximately twice the value that had been estimated previously (Treloar 1975).

It is also of interest to compare the values of r/r_m at the beginning of the upturn with some theoretical results on distribution functions for PDMS chains of finite length (Flory and Chang 1976). Of relevance here are the calculated values of r/r_m at which the Gaussian distribution function starts to overestimate the probability of extended configurations, as judged by comparisons with the results of Monte Carlo simulations. Their results are in excellent agreement with those presented here.

A second important characteristic is the value α_r of the elongation at which rupture occurs. The corresponding values of r/r_m show that rupture generally occurs at approximately 80–90% of maximum chain extensibility (Andrady et al. 1980). These quantitative results on chain dimensions are very important but may not apply directly to other networks, in which the chains could have very different configurational characteristics and in which the chain-length distribution would presumably be quite different from the very unusual bimodal distribution intentionally produced in the present networks.

IMPROVEMENTS IN ULTIMATE PROPERTIES

In this application, it is most illuminating simply to plot the nominal stress against elongation. Typical results are shown schematically in Figure 13.6 (Mark 1982a,b, 1985a). This type of representation has the advantage of having the area under each curve correspond to the network rupture energy. This energy E_r required for rupture is the standard measure of *toughness* of elastomers. In the case of the unimodal networks, E_r is relatively small. As can be seen from the figure, this is due to the small maximum extensibility in the case of all short chains, and to the small maximum values of the nominal stress in the case of all long chains. Thus, unfilled unimodal elastomers are generally very

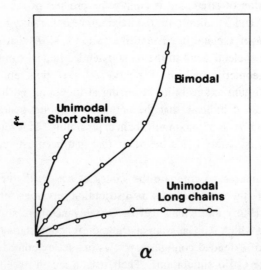

Figure 13.6 Schematic stress–strain data on unimodal and bimodal noncrystalliz-able networks plotted so that the area under each curve corresponds to the energy of rupture (Mark 1982a, 1985a).

weak materials. The bimodal networks have improved ultimate properties in that they can be prepared so as to have relatively large values of the nominal stress without the usual corresponding decrease in maximum extensibility. This short-chain reinforcing effect is very striking in that E_r can easily be increased by a factor of 5 in going from 0 to 90 mol% of the very short chains, and this corresponds to an increase of only approximately 10 wt%!

THE OBSERVED EFFECTS AS NON-GAUSSIAN BEHAVIOR

Since the results cited above demonstrate that the upturns in modulus are due to limited chain extensibility, it becomes important to interpret them in terms of a non-Gaussian theory of rubberlike elasticity. A recent novel approach to this problem utilizes the wealth of information that rotational isomeric state theory provides on the spatial configurations of chain molecules (Curro and Mark 1984; Mark 1985c). Specifically, Monte Carlo calculations based on the rotational isomeric state approximation are used to simulate spatial configurations and thus distribution functions for the end-to-end separation of the chains. These distribution functions are used in place of the Gaussian function in the standard three-chain network model (Treloar 1975) in the affine limit to give the desired non-Gaussian molecular theory of rubberlike elasticity. Stress–strain isotherms calculated in this way are strikingly similar to experimental isotherms obtained for the bimodal networks.

CRYSTALLIZABLE BIMODAL NETWORKS

In the case of the PDMS networks, the reinforcing effects were shown to be due to the limited extensibility of the very short chains. This inherent toughening effect is intramolecular and should, at least in principle, occur for bimodal networks prepared from any elastomeric polymer. It now becomes relevant to ask if there might be an *additional* effect in the case of elastomers capable of undergoing strain-induced crystallization (as described in Chapters 1 and 12). Specifically, would the very short chains in a bimodal network facilitate such crystallization, perhaps by an increased tendency to be oriented by the deformation and thus to initiate nucleation?

Comparisons of stress–strain isotherms obtained for bimodal and unimodal PDMS networks at very low temperatures could possibly answer this question. Such comparisons have been attempted, but since PDMS has such a low melting

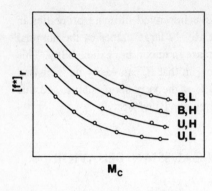

Figure 13.7 Ultimate strength shown as a function of the molecular weight between cross-links for unimodal (U) and bimodal (B) crystallizable networks at high (H) and low (L) temperatures.

point and crystallizes only sluggishly, the results were very ambiguous. More recent experiments on networks of polyoxyethylene, which has a much higher melting point (~65°C), were much more definitive (Sun and Mark 1987). The results are shown schematically in Figure 13.7. The pair of curves pertaining to the higher temperature, B, H, and U, H, demonstrate that the bimodal networks have higher values of the ultimate strength $[f^*]_r$ than the unimodal ones at the same molecular weight between cross-links. The corresponding results at a lower temperature are given by the curves B, L and U, L. Decreasing the temperature lowers the values of $[f^*]_r$ for the unimodal networks, which is consistent with the usual decrease in reduced stress with decrease in temperature, as described in Chapter 9. The values for the bimodal networks actually show an increase, presumably because of an enhanced ability to undergo strain-induced crystallization. The extent to which the bimodal networks are superior thus increases at the lower temperature. This could be of practical as well as fundamental importance.

14

BIREFRINGENCE

INTRODUCTION

The directions of chain segments are isotropically distributed in an undistorted network. Macroscopic distortion of the network, however, results in anisotropy at the microscopic level in that the distribution of segment orientations is no longer the same in all directions. The anisotropy resulting from macroscopic deformation is accompanied by optical anisotropy associated with these varying orientations of the polarizability tensors of the repeat units. Consequently the refractive indices of the network along different directions become unequal, and the network is said to be birefringent. The theory of birefringence of affine networks has been outlined by Treloar (1975). In this chapter, the derivation of birefringence will be given for both affine and phantom networks. In the following chapter, the more general problem of the segmental orientation itself will be presented.

BIREFRINGENCE OF AFFINE AND PHANTOM NETWORKS

The ends of a network chain are assumed to be fixed at the origin and the point (x, y, z) of a laboratory-fixed coordinate system $Oxyz$. The x axis is chosen to lie along the direction of stretch of the network. The segments of the chain are assumed to obtain all possible configurations, subject to the constancy of the end-to-end vector **r**.

The polarizability tensor for the chain with a given configuration is obtained as the sum of the polarizability tensors of structural units comprising the chain (Volkenstein 1963; Flory 1969). For a chain with a fixed end-to-end vector it will assume different values depending on the spatial configurations of the structural or repeat units. The difference between the x component and the two lateral components, y and z, of the polarizability tensor averaged over all possible configurations of the chain at fixed **r** (Volkenstein 1963; Nagai 1964) is given approximately by

$$\left[\alpha_{xx} - \frac{(\alpha_{yy} + \alpha_{zz})}{2} \right]_r = \Gamma_2 \frac{x^2 - (y^2 + z^2)/2}{\langle r^2 \rangle_0} \tag{14.1}$$

Here, α_{xx}, α_{yy}, and α_{zz} represent the x, y, and z components of the polarizability tensor for the chain averaged over all configurations. The subscript **r** denotes that this averaging is performed for the chain with fixed ends. On the right-hand side of Eq. (14.1) x, y, and z represent the respective components of **r**, and $\langle r^2 \rangle_0$ is the mean-squared end-to-end distance of the corresponding free chain. The quantity Γ_2 (Nagai 1964; Flory 1969) is defined by

$$\Gamma_2 = \frac{9}{10} \sum_i \frac{\langle \mathbf{r}^T \hat{\alpha}_i \mathbf{r} \rangle_0}{\langle r^2 \rangle_0} \tag{14.2}$$

where the summation is over all the structural units comprising the chain, and $\hat{\alpha}_i$ is the anisotropic part of the polarizability tensor of the ith structural unit. The quantity $\mathbf{r}^T$ is the transpose of **r** and subscript 0 denotes that the average is performed over all configurations of the free chain. Equation (14.1) may be written similarly for the other components. For a sufficiently long freely jointed chain, $\Gamma_2 = 3\Delta a/5$, where Δa is the difference in the components of the bond polarizability tensor parallel and perpendicular to the bond axis. Assuming the x, y, and z axes to be the principal directions of macroscopic deformation and averaging Eq. (14.1) over all chains of the network leads to

$$\langle \alpha_{xx} - (\alpha_{yy} + \alpha_{zz})/2 \rangle = \nu^{-1} \sum_{i=1}^{\nu} [\alpha_{xx} - (\alpha_{yy} + \alpha_{zz})/2]$$

$$= \Gamma_2 \frac{\langle x^2 \rangle - (\langle y^2 \rangle + \langle z^2 \rangle)/2}{\langle r^2 \rangle_0} \tag{14.3}$$

$$= (\Gamma_2/3) [\Lambda_x^2 - (\Lambda_y^2 + \Lambda_z^2)/2]$$

Here, the second line follows by the use of Eq. (14.1). The symbols Λ_x^2, Λ_y^2, and Λ_z^2 in the third line of Eq. (14.3) are the components of the microscopic deformation tensor (Erman and Flory 1983) defined as

$$\Lambda_x^2 = \frac{\langle x^2 \rangle}{\langle x^2 \rangle_0}, \qquad \Lambda_y^2 = \frac{\langle y^2 \rangle}{\langle y^2 \rangle_0}, \qquad \Lambda_z^2 = \frac{\langle z^2 \rangle}{\langle z^2 \rangle_0} \qquad (14.4)$$

According to the Lorentz–Lorenz relation, the refractive index difference $\Delta n \equiv n_x - (n_y + n_z)/2$ along the x axis and the y and z axes is

$$\Delta n = (2\pi/9)\,(\nu/V)\,[(n^2 + 2)^2/n]\,\langle \alpha_{xx} - (\alpha_{yy} + \alpha_{zz})/2 \rangle \qquad (14.5)$$

where n is the mean refractive index and V is the volume of the network during the experiment. Using Eq. (14.3) in Eq. (14.5) leads to

$$\Delta n = (2\pi/27)\,(\nu/V)\,[(n^2 + 2)^2 \Gamma_2/n]\,[\Lambda_x^2 - (\Lambda_y^2 + \Lambda_z^2)/2] \qquad (14.6)$$

For an affine network, Λ_x^2, Λ_y^2, and Λ_z^2 are obtained from Eqs. (4.10) and (14.4), as

$$\Lambda_x^2 = \lambda_x^2, \qquad \Lambda_y^2 = \lambda_y^2, \qquad \Lambda_z^2 = \lambda_z^2 \qquad (14.7)$$

where λ_x^2, λ_y^2, and λ_z^2 are the components of the macroscopic deformation tensor along the x, y, and z axes, respectively. Substituting Eq. (14.7) in Eq. (14.6) leads to the expression for the birefringence of an affine network,

$$(\Delta n)_{\mathrm{af}} = (\nu/V)kTC[\lambda_x^2 - (\lambda_y^2 + \lambda_z^2)/2] \qquad (14.8)$$

where

$$C = 2\pi(n^2 + 2)^2 \Gamma_2/27nkT \qquad (14.9)$$

For a phantom network, the components Λ_x^2, Λ_y^2, and Λ_z^2 are obtained from Eq. (4.17) as

$$\Lambda_x^2 = (1 - 2/\phi)\lambda_x^2 + 2/\phi$$
$$\Lambda_y^2 = (1 - 2/\phi)\lambda_y^2 + 2/\phi \qquad (14.10)$$
$$\Lambda_z^2 = (1 - 2/\phi)\lambda_z^2 + 2/\phi$$

Using Eq. (14.10) in Eq. (14.6) leads to the birefringence of a phantom network:

$$(\Delta n)_{ph} = (\xi/V)kTC[\lambda_x^2 - (\lambda_y^2 + \lambda_z^2)/2] \qquad (14.11)$$

The expression for birefringence may be written in general as

$$\Delta n = 2(\mathcal{F}kT/V)C[\lambda_x^2 - (\lambda_y^2 + \lambda_z^2)/2] \qquad (14.12)$$

where $\mathcal{F} = \nu/2$ for an affine network and $\xi/2$ for a phantom network.

Most birefringence experiments are performed in uniaxial extension for which the state of deformation is given by Eqs. (6.9a,b). Substituting Eqs. (6.9a,b) in Eq. (14.12) leads to the birefringence for uniaxial extension:

$$\Delta n = 2(\mathcal{F}kT/V_0)(V/V_0)^{-1/3}C(\alpha^2 - \alpha^{-1}) \qquad (14.13)$$

According to the theory of strain birefringence developed by Erman and Flory (1983), the birefringence of real networks should lie between that of the affine and phantom network models. Presence of short-range intermolecular orientational correlations in the network, however, leads to a significant increase of birefringence over the theoretically predicted values.

The true stress τ in simple tension is given as

$$\tau = 2(\mathcal{F}kT/V_0)(V/V_0)^{-1/3}(\alpha^2 - \alpha^{-1}) \qquad (14.14)$$

as was shown in Chapter 6. Dividing Eq. (14.13) by Eq. (14.14) leads to

$$\Delta n/\tau = C \qquad (14.15)$$

The constant C is referred to in the literature as the *stress-optical coefficient* (Treloar 1975).

SOME USES OF BIREFRINGENCE MEASUREMENTS

The most fundamental use of birefringence measurements is to evaluate the molecular theory giving rise to the equations in the preceding section. More specifically, the equations may be tested by measurements of Δn as a function of strain and temperature for networks having different degrees of cross-linking (with the cross-links possibly introduced at different reference volumes V_0) and different degrees of swelling (frequently with solvents of varying geometric asymmetry). The experimental results, some of which have been summarized by Stein (1976), are generally in good agreement with theory.

This testing of the Gaussian theories can be extended to non-Gaussian theories, particularly by the use of bimodal PDMS elastomers having very high deformability (Galiatsatos and Mark 1987). Measurements of Δn as a function of α are very useful, for example, in locating the onset of non-Gaussian behavior. Thus the birefringence data can supplement the more extensive mechanical property data (Treloar 1975) in the evaluation of non-Gaussian theories in general.

Since birefringence is very sensitive to ordering of any type, it can also be used to determine small amounts of crystallinity (Stein 1976). Specific examples are the detection of small amounts of strain-induced crystallization in networks of highly amorphous ethylene–propylene copolymers (Llorente and Mark 1981) and stereochemically irregular polybutadiene (Mark and Llorente 1981). Also relevant in this regard was the (unsuccessful) search for crystallinity in bimodal PDMS networks over a wide range in temperature (Zhang and Mark 1982). Bimodal networks were discussed in Chapter 13.

A related application is the study of the short-range correlations occurring between polymer segments and diluent, particularly as a function of the geometric asymmetry of the diluent molecule (Treloar 1975; Gent 1969; Llorente et al. 1983). An important result (Liberman et al. 1972; Liberman et al. 1974) is the finding that use of a swelling diluent that is as nearly spherical as possible gives values of Δn that are most suitable for interpretation in terms of rotational isomeric state theory (Flory 1969).

In such theoretical analyses, properties related to the birefringence are compared with corresponding results calculated using a rotational isomeric state model of the network chains. These comparisons between experiment and theory then yield chain conformational energies that can be used to predict other configuration-dependent properties. Alternatively, if the conformational energies are already known, then it is possible to obtain polarizabilities and optical anisotropies useful in the prediction of other optical properties.

15

SEGMENTAL ORIENTATION

INTRODUCTION

Strain birefringence, presented in Chapter 14, is a measure of the orientation of the polarizability tensors associated with structural units of the chains and may therefore be adopted as an average indication of segmental orientation. It should be noted, however, that birefringence is only a crude qualitative estimate of such orientation because of uncertainties in the polarizability tensor resulting from intra- and intermolecular effects and solvent–chain interactions. Recent developments in spectroscopic techniques (Ward 1975) allow more accurate measurement of orientations of specific vector directions within chain segments. For example, infrared dichroism, ^{2}H NMR spectroscopy, and fluorescence polarization are three of the more widely used methods.

RELATION OF SEGMENTAL ORIENTATION TO DEFORMATION

The degree of orientation of segments in a deformed network is given usually in terms of the orientation function S defined by

$$S = (3\langle \cos^2 \theta \rangle - 1)/2 \qquad (15.1)$$

117

Here $\langle \cos^2 \theta \rangle$ represents the mean-squared projection of the segment vectors on a laboratory-fixed axis, where θ is the angle between the segment direction and the axis. If all segments are directed along the given axis, $\langle \cos^2 \theta \rangle = 1$ and $S = 1$. In the case where all segments are perpendicular to the given axis, $\langle \cos^2 \theta \rangle = 0$ and $S = -\frac{1}{2}$. For random orientation of segments, $\langle \cos^2 \theta \rangle = \frac{1}{3}$ and $S = 0$.

Evaluation of S in terms of molecular parameters for a given network is made by first considering a network chain with fixed end-to-end vector $\mathbf{r}$ as shown in Figure 15.1. In this figure, $0xyz$ is a laboratory-fixed coordinate system. The sample is assumed to be stretched (or compressed) along the x axis; y and z axes are lateral directions. The vector $\mathbf{m}_i$ is affixed to the ith structural unit of the chain. It may be associated with a fictitious direction in the unit or may be identified with an actual bond. The angle between $\mathbf{m}_i$ and the x axis (not shown in the figure) is θ. As the chain undergoes configurational transitions, subject to the constancy of $\mathbf{r}$, the angle θ takes different values. The average $\overline{\cos^2 \theta}$ over all configurations of the chain is given according to the statistical treatment (Nagai 1964; Flory 1969) as

$$\overline{\cos^2 \theta} = \frac{1}{3}\left\{ 1 + 2D_0\left[\frac{x^2}{\langle x^2 \rangle_0} - \frac{1}{2}\left(\frac{y^2}{\langle y^2 \rangle_0} + \frac{z^2}{\langle z^2 \rangle_0} \right) \right] \right\} \tag{15.2}$$

where

$$D_0 = \frac{3\langle r^2 \cos^2 \Phi \rangle_0 / \langle r^2 \rangle_0 - 1}{10} \tag{15.3}$$

with Φ indicating the angle between $\mathbf{m}_i$ and the chain vector $\mathbf{r}$. The subscript zero in Eqs. (15.2) and (15.3) indicates that the averaging is performed on free

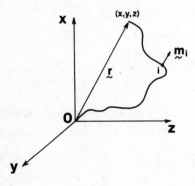

Figure 15.1 Schematic representation of a chain with fixed end-to-end vector $\mathbf{r}$, and a vector $\mathbf{m}_i$ affixed to unit i of the chain.

chains. The average $\overline{\cos^2 \theta}$ given by Eq. (15.2) indicates an average over all configurations of the chain with one end at the origin and the other kept fixed at point (x, y, z). Averaging Eq. (15.2) over all chains of the network leads to

$$\langle \cos^2 \theta \rangle = \frac{1}{3}\left\{1 + 2D_0\left[\Lambda_x^2 - \frac{1}{2}(\Lambda_y^2 + \Lambda_z^2)\right]\right\} \qquad (15.4)$$

where $\Lambda_x^2 = \langle x^2 \rangle / \langle x^2 \rangle_0$, $\Lambda_y^2 = \langle y^2 \rangle / \langle y^2 \rangle_0$, and $\Lambda_z^2 = \langle z^2 \rangle / \langle z^2 \rangle_0$ denote the components of the molecular deformation tensor given in Chapter 14. Substituting Eq. (15.4) into Eq. (15.1) leads to the orientation function in terms of the components of the molecular deformation tensor and the molecular quantity D_0:

$$S = D_0[\Lambda_x^2 - (\Lambda_y^2 + \Lambda_z^2)/2] \qquad (15.5)$$

The derivation of Eq. (15.5) is based on the assumption of Gaussian chains that are sufficiently below their finite extensibility limits. At higher deformations, the approximation given by Eq. (15.2) is not sufficient and additional terms are required.

Components of the molecular deformation tensor are given in terms of macroscopic deformation for an affine and a phantom network by Eqs. (14.7) and (14.10), respectively. Thus the orientation function for an affine network follows by substituting Eq. (14.7) into Eq. (15.5) to get

$$S = D_0[\lambda_x^2 - (\lambda_y^2 + \lambda_z^2)/2] \qquad (15.6)$$

Similarly, for a phantom network, using Eq. (14.10) in Eq. (15.5) leads to

$$S = D_0\left(1 - \frac{2}{\phi}\right)\left(\lambda_x^2 - \frac{1}{2}(\lambda_y^2 + \lambda_z^2)\right) \qquad (15.7)$$

Simple extension is the most widely used deformation for measurements of segmental orientation. Using $\lambda_x = (v_{2C}/v_2)^{1/3}\alpha$ and $\lambda_y = \lambda_z = (v_{2C}/v_2)^{1/3} \alpha^{-1/2}$ for simple extension in Eqs. (15.6) and (15.7) leads to

$$S = D_0\left(\frac{v_{2C}}{v_2}\right)^{2/3}(\alpha^2 - \alpha^{-1}) \quad \text{(affine)} \qquad (15.8)$$

and

$$S = D_0\left(1 - \frac{2}{\phi}\right)\left(\frac{v_{2C}}{v_2}\right)^{2/3}(\alpha^2 - \alpha^{-1}) \quad \text{(phantom)} \qquad (15.9)$$

By analogy with the reduced stress $[f^*]$, a reduced orientation function may be defined, by using Eq. (15.9) for the phantom network, as

$$[S] \equiv \frac{S}{(1 - 2/\phi)(v_{2C}/v_2)^{2/3}(\alpha^2 - \alpha^{-1})}$$

$$= D_0 \qquad (15.10)$$

The second line of Eq. (15.10) suggests that the reduced orientation function should be independent of deformation, as was the case for the reduced stress of a phantom network. Experiments show, however, that $[S]$ depends rather strongly on α. Typical results of experiments (Queslel et al. 1985) are shown in Figure 15.2. The circles represent experimental data, for dry and swollen samples. The horizontal dashed line, representing the reduced orientation defined in Eq. (15.10) is given an arbitrary location along the ordinate in the figure. The solid curves are obtained according to the constrained junction theory (Erman and Monnerie 1985; Queslel et al. 1985). The strong dependence of $[S]$ on swelling shown by the data points indicates the presence of short-range intermolecular orientational correlations that diminish rapidly upon swelling. The original theory on segmental orientation (Kuhn and Grün 1942) is based on an affine network with freely jointed chains under simple tension. For this case, with $v_{2C} = v_2 = 1$, Eq. (15.8) reduces to

$$S = \tfrac{1}{5}N(\alpha^2 - \alpha^{-1}) \qquad (15.11)$$

where N is the number of links in the chain. In comparing predictions of Eq. (15.11) with experimental data on networks whose chains are not freely jointed, the number of bonds N has to be reinterpreted as the number of Kuhn segments in the chain.

The expressions for S given by Eqs. (15.8) and (15.9) for the Affine and Phantom Network models reflect segmental orientation in the absence of inter-molecular orientational correlations. According to this treatment, Eqs. (15.8) and (15.9) constitute upper and lower bounds to S. Intermolecular orientational correlations in real networks result in larger values of S. Dilation of the network with an isotropic solvent leads to rapid removal of such correlations, however,

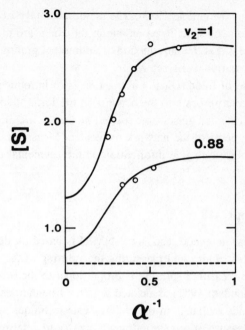

Figure 15.2 Reduced orientation [S] as a function of the reciprocal extension ratio α^{-1}. Circles are from experiments of Queslel et al. (1985) on dry ($v_2 = 1$) and swollen ($v_2 = 0.88$) networks, where v_2 is the value of the volume fraction of polymer for each curve. The horizontal dashed line represents the prediction of the elementary molecular theory, and the solid curves are from the constrained junction theory.

as mentioned above. Identification of the various contributions to segmental orientation seems to be of major importance for correct interpretation of experimental data in this area.

EXPERIMENTAL DETERMINATION OF SEGMENTAL ORIENTATION

(a) Fluorescence Polarization

Fluorescence polarization is based on labeling a small fraction of the network chains by fluorescent molecules, which absorb and emit light along specific directions. The orientations of these fluorescent groups may then be determined by exciting them with polarized light and measuring the emitted intensity along different directions (Monnerie 1983). The major advantage of this technique is that the orientations of specific points along the chain are measured directly. One disadvantage is that the introduction of fluorescent groups into a chain may perturb its configurational character.

A different use of fluorescence polarization is to introduce structurally anisotropic fluorescent probes into the network in the form of solvent molecules. Experiments indicate that anisotropic solvent molecules are orientationally coupled with the segments of the network molecules. Measurements of the orientations of the probes thus give information on the segmental orientation in the networks.

(b)   Deuterium NMR

Deuterium nuclear magnetic resonance (NMR) is based on deuterating a fraction of the network chains and measuring the splitting of the NMR lines. The magnitude of the measured splitting is proportional to the orientation function (Deloche and Samulski 1981) associated with the orientations of the hydrogen or deuterium bonds or labels in the chains. This technique seems to give the most sensitive measure of orientation, enabling one to determine minute levels of orientation in deformed networks. As in fluorescence polarization, the deuterium NMR technique may be used for measuring the orientations of deuterated solvent molecules, or probes, that are introduced into the network by swelling.

(c)   Polarized Infrared Spectroscopy

This technique is based on the measurement of orientations of specific transition dipole moment directions having known infrared absorption bands. It is based on the determination of the dichroic ratio for a deformed network. For a network under simple tension, this ratio is defined as

$$D = A_\parallel / A_\perp \qquad (15.12)$$

where $A_\parallel$ and $A_\perp$ are the absorbances of radiation polarized along and perpendicular to the direction of stretch, respectively. For an isotropic, unoriented network, the dichroic ratio equals unity, and the deviation of D from unity is proportional to the orientation function (Read and Stein 1968; Flory 1969). Calculations based on the rotational isomeric state theory provide the relationship between the experimentally measured value of the dichroic ratio and the desired orientation function.

16

ROTATIONAL ISOMERIZATION

INTRODUCTION

As was first mentioned in Chapter 1, one of the distinguishing characteristics of a flexible polymer chain is its large deformability. This feature depends mainly on the possibility of rotations about the backbone bonds of the chain. These rotations are subject to energy barriers, as illustrated for the ith bond of a chain in Figure 16.1. According to the rotational isomeric state theory, the bond rotations are assumed to obtain only the angles corresponding to the energy minima, identified by the angles $\phi_i(t)$, $\phi_i(g^+)$, and $\phi_i(g^-)$ in Figure 16.1a. The letters t, g^+, and g^- refer to the trans, gauche$^+$, and gauche$^-$ states of the bond, respectively (Flory 1969). The three-minima map of Figure 16.1b is representative of several polymeric species. Other maps may be simpler or more complicated depending on the molecular structure of the chain.

A chain fixed at two ends assumes various configurations through rotations about its skeletal bonds. The average numbers of t, g^+, and g^- states for the given fixed end-to-end vector $\mathbf{r}$ characterizes the configurational energy of the chain. The average number of rotational isomeric states is expected to change upon modifying the end-to-end vector $\mathbf{r}$. This change is referred to as *rotational isomerization of polymer chains by stretching*.

Methods for calculating the populations of rotational isomeric states were developed by Volkenstein (1963), Birshtein and Ptitsyn (1966), and Flory

123

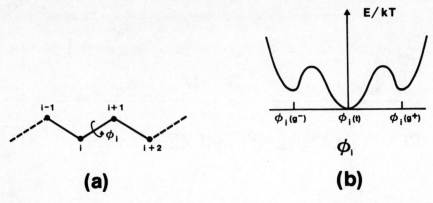

Figure 16.1 (a) A few backbone bonds of a chain, where ϕ_i is the rotation around the ith bond (which joins skeletal atoms i and $i + 1$). (b) The potential for rotations ϕ_i. The minima in the energy are identified as gauche⁻ (g⁻), trans (t), and gauche⁺ (g⁺). Corresponding values of ϕ_i are indicated along the abscissa as $\phi_i(g^-)$, $\phi_i(t)$, and $\phi_i(g^+)$.

(1969). The problem was formulated by assuming that the chains are sufficiently long and are well below their maximum extensions even in the deformed state. Calculations based on these two assumptions showed that stretching has very little effect on rotational isomerization. It must be noted, however, that the chains in a real network are often much shorter and are subject to much larger extensions than the assumptions of the theory allow for. Recent developments in infrared spectroscopy and its growing application to analysis of elastomeric networks place renewed interest in rotational isomerization by stretching. We will therefore briefly cover the mathematical treatment of this subject in this chapter. The formulation is based on the work of Abe and Flory (1970).

A SIMPLIFIED DESCRIPTION OF THE THEORY

Formulation of the problem is based on calculating the change of the average number of bonds in a given state when the end-to-end vector of a free chain is constrained to a fixed value r. According to the theory, this change is given by

$$\overline{\Delta n}_{\eta;r} = \overline{n}_{\eta;r} - \overline{n}_{\eta}$$

$$= D_2\left(\frac{r^2}{\langle r^2 \rangle_0} - 1\right) \tag{16.1}$$

Here, $\bar{n}_{\eta;r}$ represents the average number of bonds in state η when the chain is kept at fixed end-to-end length r. $\bar{n}_\eta$ is the corresponding number for the same state when the chain ends are free. $\overline{\Delta n_{\eta;r}}$ thus represents the change in the number of bonds in state η when the end-to-end length of the free chain is constrained to the value r. D_2 in Eq. (16.1) is the coefficient of proportionality that relates the change $\overline{\Delta n_{\eta;r}}$ to the state of deformation of the single chain described by the term in the brackets. It may be calculated for a chain of given structure by the matrix generation technique (Flory 1969) of the rotational isomeric state formalism. The averages indicated by the overbars in Eq. (16.1) represent time averages over all configurations of the single chain subject to the stated constraint. Summing both sides of Eq. (16.1) over all ν chains of the network leads to

$$(\Delta n_\eta)_{V,\alpha} = \nu D_2\left(\frac{\langle r^2 \rangle}{\langle r^2 \rangle_0} - 1\right) \tag{16.2}$$

Here the angular brackets represent an average over all chains of the network, and the subscripts V,α indicate that the volume and the state of macroscopic deformation are kept constant during averaging. The ratio $\langle r^2 \rangle / \langle r^2 \rangle_0$ in Eq. (16.2) is the molecular deformation tensor defined in Chapter 14. Equation (16.2) may be written in terms of x, y, z components:

$$\langle \Delta n_\eta \rangle_{V,\alpha} = \nu D_2\left(\frac{\Lambda_x^2 + \Lambda_y^2 + \Lambda_z^2}{3} - 1\right) \tag{16.3}$$

For simple tension these components Λ_x^2, Λ_y^2, and Λ_z^2 are given for affine and phantom networks by Eqs. (14.7) and (14.10), respectively. Substituting these equations into Eq. (16.3) leads to

$$\langle \Delta n_\eta \rangle_{V,\alpha} = \frac{2}{3}\, \mathfrak{F}D_2\left(\frac{v_{2C}}{v_2}\right)^{2/3} (\alpha^2 + 2\alpha^{-1} - 3) \tag{16.4}$$

The state η as defined above represents the conformational state of a single bond or a group of bonds along the chain. Calculations of Abe and Flory (1970) show that the value of D_2 changes very little with chain end-to-end length for long (amorphous) polyethylene chains. Similar conclusions have been reached upon analysis of experimental isomerization studies of polyisoprene (Abe and Flory 1971a) and polybutadiene (Abe and Flory 1971b).

Table 16.1 Rotational Isomerization in an Affine Network at Elongations of 2 and 6

Rotational Conformation η	$D_2{}^a$	$\frac{1}{3}D_2(\alpha^2 + 2\alpha^{-1} - 3)$	
		$\alpha = 2$	$\alpha = 6$
t	0.655	0.87	7.71
g$^\pm$	−0.328	−0.44	−3.86
tt	0.764	1.02	9.00
g$^\pm$g$^\pm$	−0.214	−0.29	−2.52
ttt	0.884	1.18	10.41
tg$^\pm$g$^\pm$	−0.185	−0.25	−2.18
tttt	0.650	0.87	7.66
g$^\pm$ttg$^+$	−0.151	−0.20	−1.78

^{a}Values from Abe and Flory (1970).

Results of calculating $\langle \Delta n \rangle_{V,\alpha}$ for a network of polyethylene chains having 1027 skeletal bonds are shown in Table 16.1. The conformational state of a bond or group of successive bonds is shown in the first column. The values of D_2, shown in the second column, are taken from the work of Abe and Flory (1970). Calculations are made for $\alpha = 2$ and 6, and for $v_{2C} = v_2 = 1$. The third and fourth columns indicate the change in the average number of bonds in state η per chain of an affine network. This is represented by $\frac{1}{3}D_2(\alpha^2 + 2\alpha^{-1} - 3)$. The entries in the table show that, upon stretching, the number of ttt sequences rapidly increases and the number of g$^\pm$g$^\pm$ sequences decreases. Dependence on the level of strain is strong due to the α^2 term in Eq. (16.4). However, predictions of the theory for $\alpha = 6$ should be used with caution because the approximations involved in the derivations are based on chains that are well below their maximum extensions.

Predictions of the theory developed above are expected to help the understanding of strain-induced crystallization in networks. The tendency of a network to crystallize upon stretching depends strongly on the increase in the number and length of regular, crystallizable sequences, such as the ttt sequence described in Table 16.1.

17

OSMOTIC COMPRESSIBILITY, CRITICAL PHENOMENA, AND GEL COLLAPSE

INTRODUCTION

A swollen network exposed to the continuous action of a solvent forms a *semiopen thermodynamic system* where only the solvent molecules may enter or leave the swollen network. Recent experimental evidence has shown that the extent to which the solvent enters or leaves the system may be controlled by changing variables such as temperature, solvent activity, etc. Certain network–solvent systems have shown that the change in volume due to the motion of the solvent may be very abrupt and significant and resemble a first-order transition. This phenomenon is now referred to as *gel collapse*, where the word *gel* being used synonymously with swollen network. In this chapter, we review the thermodynamics of networks in equilibrium with a solvent and discuss the nature of the observed transitions.

CHEMICAL POTENTIAL OF THE SOLVENT IN A SWOLLEN NETWORK

The state of equilibrium swelling of a network immersed in a solvent is obtained when the solvent inside the network is in thermodynamic equilibrium with the solvent outside (Flory 1953; Treloar 1975; Queslel and Mark 1985a). This state of equilibrium is described by the thermodynamic relation

$$\Delta\mu_1 \equiv \mu_1 - \mu_1^o$$

$$= \left(\frac{\partial \Delta A}{\partial n_1}\right)_{T,p} = RT \ln a_1 \tag{17.1}$$

where $\Delta\mu_1$ is the difference in the solvent chemical potential inside (μ_1) and outside (μ_1^o) the swollen network, a_1 is the activity of the surrounding solvent, and ΔA is the Gibbs free energy of the swollen network. The term $(\partial \Delta A/\partial n_1)_{T,p}$ is described by Eq. (7.6) in terms of the mixing and the elastic free energies. For a phantom network, which satisfactorily represents the behavior of swollen real networks as stated in Chapter 7, this term is given by Eq. (7.8). Substituting Eq. (7.8) into Eq. (17.1) leads to

$$\frac{\Delta\mu_1}{RT} = \ln\left(1 - v_{2m}\right) + \chi v_{2m}^2 + v_{2m} + B\left(\frac{v_{2m}}{v_{2C}}\right)^{1/3} = \ln a_1 \tag{17.2}$$

The behavior of a network at any degree of swelling is governed by Eq. (17.2). The polymer–solvent interaction parameter χ (Flory 1953) in general is dependent on both temperature and solvent. It may be expressed in a Taylor series as

$$\chi = \chi_1(T) + \chi_2(T)v_{2m} + \chi_3(T)v_{2m}^2 + \ldots \tag{17.3}$$

where χ_1, χ_2, etc. are functions of temperature only. Neglecting second- and higher-order terms in v_{2m} in this series expansion is a satisfactory approximation for a large group of polymer–solvent systems.

The state of *equilibrium swelling* depends on temperature and solvent activity according to Eq. (17.2). Changing either one or both of these two parameters results in a change of the degree of swelling occurring at equilibrium. Depending on the constitution of the network and the χ parameter, the extent of swelling and deswelling may be significantly large. Such large variations in swelling are of great importance for biological as well as synthetic network–solvent systems.

OSMOTIC COMPRESSIBILITY

The osmotic pressure π of a swollen network is defined as

$$\pi = -\Delta\mu_1/V_1 \tag{17.4}$$

The osmotic compressibility κ_{os}, which reflects the change in the degree of swelling due to a change in the osmotic pressure, is defined as

$$\kappa_{os} = v_{2m}^{-1}\left(\frac{\partial \pi}{\partial v_2}\right)_{T,p}^{-1} \tag{17.5}$$

Substituting Eq. (17.4) into Eq. (17.5) and using Eq. (17.2) (Bahar and Erman 1987) leads to

$$\kappa_{os}^{-1} = K_v = \left(\frac{RT}{V_1}\right)\left(\frac{v_{2m}^2}{1 - v_{2m}} - 2\chi_1 v_{2m}^2 - 3\chi_2 v_{2m}^3 - \left(\frac{B}{3}\right)\left(\frac{v_{2c}}{v_{2m}}\right)^{-1/3}\right)$$

$$(17.6)$$

Here K_v is the bulk osmotic modulus at fixed temperature and pressure. Careful measurements of osmotic compressibility on various systems by Zrinyi and Horkay (1982) support the validity of this relationship. Equation (17.6) shows that the magnitude of the osmotic compressibility depends strongly on the equilibrium degree of swelling, which is obtained by solving Eq. (17.2) for v_{2m}. Thus, according to Eqs. (17.2) and (17.6), κ_{os} is a function of temperature

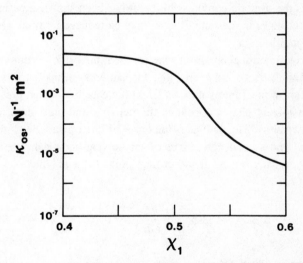

Figure 17.1 Osmotic compressibility κ_{os} as a function of χ_1 for a typical gel (Bahar and Erman 1987). The curve shows results of calculations obtained from Eqs. (17.2) and (17.6).

(through χ_1 and χ_2) and of chemical activity a_1 of the surrounding solvent. In general, the osmotic compressibility is high for good solvents and decreases progressively as the solvent becomes poorer. Results of calculations for a typical network–solvent system are shown in Figure 17.1, where osmotic compressibility is plotted as a function of χ_1 (χ_2 is assumed to be zero). The curve indicates that osmotic compressibility changes by more than three orders of magnitude when χ_1 varies over only the small interval 0.4–0.6.

The occurrence of $\partial \pi / \partial v_2$ in the denominator of Eq. (17.5) suggests the possibility of obtaining infinite osmotic compressibility when the former equates to zero. Experiments of Tanaka (1978) and Ilavsky (1982) have indeed shown that conditions in network–solvent systems may obtain to yield very large osmotic compressibility values. The problem of infinite osmotic compressibility is related to the more general and important problem of critical phenomena in swollen networks, which are treated in more detail in the next section.

CRITICAL PHENOMENA AND GEL COLLAPSE

The osmotic compressibility or the chemical potential of a network depends on temperature and solvent activity. If the network chains carry ionic groups, then the pH or ionic strength of the solvent, or an electric field acting across the gel, also affects the osmotic compressibility. It has been shown experimentally that very large changes in the state of a gel may be induced by small changes in the external conditions.

When certain critical conditions are reached, the osmotic compressibility becomes infinitely large indicating that the network shrinks strikingly and can continue almost indefinitely under a slight increase in external pressure. At this state, pure-solvent phases appear in the network and their sizes reach macroscopic dimensions. This indicates the onset of large fluctuations in the system.

The first and second derivatives of the solvent chemical potential with respect to v_2 equate to zero at the critical point for a polymer–solvent system. Thus,

$$\Delta \mu_1 = RT \ln a_1, \qquad \frac{\partial \Delta \mu_1}{\partial v_2} = 0, \qquad \frac{\partial^2 \Delta \mu_1}{\partial v_2^2} = 0 \qquad (17.7)$$

For a given polymer–solvent system, that is, for given χ_1, χ_2, and B, Eqs. (17.7) should hold in order for critical conditions to obtain (Erman and Flory 1986). Vanishing of the second of Eqs. (17.7) is the condition that leads to infinite osmotic compressibility, as seen from the definition of κ_{os}.

A network–solvent system may also exhibit phase separation where three phases of different concentrations may exist in equilibrium inside the swollen network. One of the phases is pure solvent ($v_2 = 0$). The remaining two phases are a less swollen and a highly swollen phase. Such a state is referred to as *triphasic equilibrium*, described by the equations

$$\Delta\mu_1(v_2') = \Delta\mu_1(v_2'') = RT \ln a_1$$

$$\int \Delta\mu_1 \, dn_1 = 0 \tag{17.8}$$

where v_2' and v_2'' represent the two coexisting polymer concentrations. This state may be reached by varying the temperature, which in turn varies $\chi_1(T)$ and $\chi_2(T)$. Alternatively, triphasic conditions may be obtained by varying the activity or the pH of the solvent in which the gel is immersed. In either case, the values of v_2' and v_2'' may be obtained by substituting Eq. (17.2) into Eq. (17.8) and solving the resulting equations. Results of typical calculations are shown in Figure 17.2 for triphasic equilibrium obtained by varying the temperature.

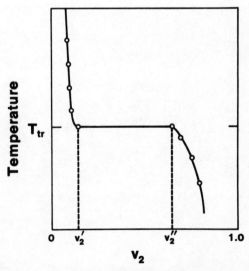

Figure 17.2 Temperature–concentration diagram for a swollen network showing a transition. T_{tr} indicates the temperature of transition at which triphasic equilibrium obtains. The quantities v_2' and v_2'' denote polymer volume fractions in the less swollen and the more highly swollen phases at this state, respectively. Circles show representative data from corresponding experiments.

The horizontal line identifies the temperature at which the transition takes place. At higher temperatures, $v_2 < v_2'$ and the network is highly swollen. At lower temperatures, $v_2 > v_2''$ and the network is collapsed. The circles are representative of experimental data obtained by various groups of researchers (Tanaka 1978; Ilavsky 1982). The width of the horizontal line at the transition depends on various parameters such as the presence of ionic groups on the network chains, the degree of cross-linking, and the amount of solvent during cross-linking. A review of critical phenomena and gel collapse has been given by Tanaka (1981).

18

NEUTRON SCATTERING FROM NETWORKS

INTRODUCTION

One of the first relevant applications of neutron scattering measurements confirmed that chains in the bulk, undeformed amorphous state exhibit their unperturbed dimensions (Flory 1984), as we mentioned in Chapter 1. Of much greater present interest, however, are measurements on deformed networks, which we describe here.

The relationship between an externally applied state of deformation and the corresponding changes at the molecular level is essentially the most important problem in the molecular theory of rubber elasticity. The phantom and the affine network models make different assumptions for this necessary relationship between the macroscopic and the microscopic states of deformation. An experimental test of these relations is possible by the *small-angle neutron scattering* (SANS) technique. According to this technique, the shape and size of network chains in the undeformed and deformed states can be measured directly. The basic principle rests on labeling certain sites, such as the junctions or points along the chains, by replacing hydrogen atoms with deuterium and then measuring the intensity of scattering from the labeled molecules. In this way, scattering from individual molecules may be measured, whereas in the more conventional light and X-ray scattering techniques, contributions from a group of neighboring molecules cannot be separated.

SANS studies of networks were initiated by Benoit and collaborators (Benoit et al. 1976) and have generated wide interest in subsequent years (Hinckley et al. 1978; Clough et al. 1980; Beltzung et al. 1982; Bastide et al. 1984).

THEORY OF SCATTERING

The geometrical variables that describe scattering from two deuterated points i and j of a polymer chain are shown in Figure 18.1. The vector from i to j is identified as $\mathbf{r}_{ij}$, and $\mathbf{k}_0$ and $\mathbf{k}$ are the wave propagation vectors for incident and scattered neutron rays, respectively. The angle between their directions is θ. Magnitudes of $\mathbf{k}_0$ and $\mathbf{k}$ are equal, $|\mathbf{k}_0| = |\mathbf{k}| = 2\pi/\lambda$, where λ is the wavelength of the radiation. The scattering vector $\mathbf{q}$ is defined as

$$q = k - k_0 \tag{18.1}$$

where

$$|\mathbf{q}| = \frac{4\pi}{\lambda} \sin \frac{\theta}{2} \tag{18.2}$$

The two points i and j may be the junctions or points along a chain. The intensity of radiation scattered in direction θ from a collection of labeled points (Flory 1969) is

$$S(\mathbf{q}) \equiv I(\theta)/I_0$$

$$= (n + 1)^{-2} \sum_{i,j} \int \exp(i\mathbf{q} \cdot \mathbf{r}_{ij}) \, W(r_{ij}) \, d\mathbf{r}_{ij} \tag{18.3}$$

Here I_0 is the scattering intensity when $\theta = 0$, $i = \sqrt{-1}$, and n is the number of scattering centers under observation. The ratio $I(\theta)/I_0$ represented by the term $S(q)$ is referred to as the *scattering law*. The intensity given by Eq. (18.3) is averaged over all possible configurations of each pair as acknowledged by the integration. $W(r_{ij})$ indicates the probability distribution of r_{ij}, referred to as the *intramolecular pair correlation function*. When labeled chains are well separated from each other, the intensity given by Eq. (18.3) corresponds to scattering from a single chain. This constitutes the advantage of neutron scattering over all other experimental techniques.

For a chain in a network in simple tension, the scattering law given by Eq.

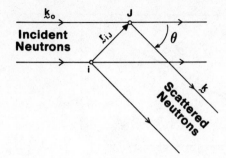

Incident Neutrons

Figure 18.1 Scattering from two points *i* and *j* separated by the vector $\mathbf{r}_{ij}$. The quantity θ is the scattering angle, and $\mathbf{k}_0$ and $\mathbf{k}$ are wave propagation vectors for incident and scattered neutrons, respectively.

(18.3) may be approximated in the long wavelength (small angle) limit (Pearson 1977) as

$$S(q_\parallel) = 1 - q_\parallel^2 \langle s^2 \rangle_\parallel$$
$$S(q_\perp) = 1 - q_\perp^2 \langle s^2 \rangle_\perp$$

(18.4)

Here, $q_\parallel$ and $q_\perp$ indicate the components of the scattering vector parallel and perpendicular to the direction of stretch, respectively. $\langle s^2 \rangle_\parallel$ and $\langle s^2 \rangle_\perp$ are the components of the mean-squared radius of gyration of the chain.

Experimental measurements of $S(q_\parallel)$ and $S(q_\perp)$ thus allow the determination of chain dimensions in a deformed network. For an affine network at uniaxial extension α, the two components of the radius of gyration are

$$\langle s^2 \rangle_\parallel = \left(\frac{v_{2C}}{v_2} \right)^{2/3} \alpha^2 \langle s^2 \rangle_{\parallel,0}$$

$$\langle s^2 \rangle_\perp = \left(\frac{v_{2C}}{v_2} \right)^{2/3} \alpha^{-1} \langle s^2 \rangle_{\perp,0}$$

(18.5)

For a phantom network, the two components of $\langle s^2 \rangle$ are calculated (Pearson 1977; Ullman 1979, 1982; Erman 1987) as

$$\langle s^2 \rangle_\parallel = \left[\left(\frac{1}{2} - \frac{1}{\phi} \right) \left(\frac{v_{2C}}{v_2} \right)^{2/3} \alpha^2 + \left(\frac{1}{\phi} + \frac{1}{2} \right) \right] \langle s^2 \rangle_{\parallel,0}$$

$$\langle s^2 \rangle_\perp = \left[\left(\frac{1}{2} - \frac{1}{\phi} \right) \left(\frac{v_{2C}}{v_2} \right)^{2/3} \alpha^{-1} + \left(\frac{1}{\phi} + \frac{1}{2} \right) \right] \langle s^2 \rangle_{\perp,0}$$

(18.6)

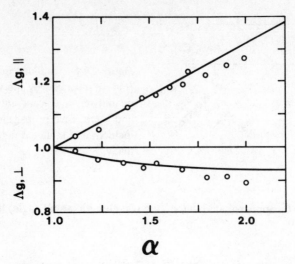

Figure 18.2 Scattering from PDMS networks (Beltzung et al. 1984). The ordinate values show the ratios, $\Lambda_\parallel = (\langle s^2 \rangle_\parallel / \langle s^2 \rangle_{\parallel,0})^{1/2}$ and $\Lambda_\perp = (\langle s^2 \rangle_\perp / \langle s^2 \rangle_{\perp,0})^{1/2}$. Curves are calculated from Eq. (18.6) for a tetrafunctional network.

Results of SANS experiments (Beltzung et al. 1984) on PDMS networks are compared with predictions of Eq. (18.6) for a tetrafunctional network in Figure 18.2. The ordinate quantities are $\Lambda_\parallel = (\langle s^2 \rangle_\parallel / \langle s^2 \rangle_{\parallel,0})^{1/2}$ and $\Lambda_\perp = (\langle s^2 \rangle_\perp / \langle s^2 \rangle_{\perp,0})^{1/2}$. Experimental data indicate that the radius of gyration in the network deforms similarly to that expected for the phantom network. There is obviously a need for more experimental studies.

19

BIOELASTOMERS

INTRODUCTION

Bioelastomers, or elastomeric biopolymers, are utilized by living organisms in a variety of tissues for a number of purposes. In vertebrates, including mammals such as man, examples of tissues are the skin, arteries and veins, and organs such as the lungs and heart. As is obvious, all these tissues involve the already-mentioned characteristics of deformability with recoverability.

There are two general reasons for studying the elasticity of such materials. The more fundamental one is the simple desire to understand rubberlike elasticity in as broad a context as possible. The more practical one is to learn how nature designs and produces these materials, so as possibly to obtain some guidance on the commercial preparation of more useful nonbiopolymeric elastomers.

The bioelastomers that have been investigated with regard to their rubberlike elasticity are listed in Table 19.1. All are proteins and thus have the repeat unit shown in Figure 19.1, where the side group R is different for the different α-amino acids that produce this chain structure. Although there are a variety of bioelastomers, elastin has been the most studied by far. It is thus the focus of the following sections.

Table 19.1   Bioelastomers Investigated with Regard to Their Rubberlike Elasticity

Bioelastomer	Occurrence
Elastin	Vertebrates
Denatured collagen[a]	Vertebrates
Resilin	Insects[b]
Abductin[c]	Mollusks[c]
Octupus arterial elastomer	Octupuses
Viscid silk	Spider webs

[a]Elastomeric only after denaturation (melting of the normally crystalline collagen fibers).
[b]Present in ligaments and tendons, particularly in jumping insects such as fleas.
[c]Present in elastic hinges to antagonize (counteract) the muscles that close the mollusk shells.

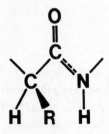

Figure 19.1   The polypeptide or protein repeat unit. (Reprinted with permission from J. E. Mark and G. Odian, Eds., *Polymer Chemistry*. Copyright American Chemical Society 1984.)

STRUCTURAL CHOICES

Elastin is, of course, a chemical copolymer, and its repeat unit sequence is sufficiently irregular that it is always totally amorphous. This use of copolymerization to suppress the large amounts of crystallinity that would interfere with elastomeric behavior is also practiced by synthetic polymer chemists. For example, introduction of some propylene units into what would otherwise be highly crystalline polyethylene results in elastomeric ethylene–propylene copolymers that are of considerable commercial importance (Borg 1973). Similarly, copolymerization of fluorocarbon monomers (which would separately yield highly crystalline homopolymers) is used to prepare a number of high-performance fluoroelastomers (Brydson 1978).

The chain flexibility required for rubberlike elasticity is allowed for in elastin through use of R side groups that are generally very small. The relevant information for mammalian elastin is summarized in the first three rows of Table 19.2 (Gosline 1980, 1987). Specifically, three unusually small side groups make up a total of 68 mol% of those occurring in this type of elastin. An analogous avoidance of numerous large groups can be seen in the structures of the typical nonbiopolymeric elastomers described in Table 2.1. It is also interesting to note, as shown in the final row of Table 19.2, that nonpolar units predominate, making elastin one of the most hydrophobic proteins known. This was thought to illustrate nature's attempt to avoid strong intermolecular interactions, such as coulombic attractions, since they could interfere with desired chain mobility. This seems not to be the case, however, judging from two additional pieces of information (Gosline 1987). First, some bioelastomers are significantly more polar than elastin. Second, the elastin chains would in any case interact only weakly, since they are well separated by the solvent molecules, which swell elastin as it functions in the organism. (The volume fraction v_{2m} of the polymer in elastin at swelling equilibrium is only approximately 0.6.) This swelling or *plasticization* with aqueous body fluids is absolutely crucial to the functioning of elastin since the dry polymer has a glass transition temperature T_g of approximately 200°C. An analogous example is the already-cited use of typical ester plasticizers to increase the pliability of poly(vinyl chloride) (Davis 1973), which otherwise has a T_g of 82°C.

The preponderance of nonpolar side groups, possibly sheathing the polar elastin backbone, may have a completely different purpose. It could, for example, be the origin of the highly unusual negative temperature coefficient of swelling that elastin exhibits in aqueous fluids. This permits the survival of fish whose body temperatures decrease when they swim into colder water. The decrease in temperature increases the degree of swelling of the elastin, thus further decreasing its T_g and keeping the chains flexible at the lowered temperature (Gosline 1987). The usual positive temperature coefficient of swelling observed for elastomers would have disastrous consequences for the fish.

Table 19.2   Composition of Mammalian Elastin

Amino Acid	R side group[a]	Mole Percent
Glycine	H	31
Alanine	CH_3	24
Valine	$CH(CH_3)_2$	13
Various	Nonpolar	61

[a]In the [CHR—CO—NH—] repeat unit.

Source: Gosline (1980, 1987).

CROSS-LINKING

Elastin chains are cross-linked in a highly specific manner, very unlike the usual techniques used to cure commercial elastomers (Gosline 1980). The cross-linking occurs through lysine repeat units, the number and placement of which along the chains are carefully controlled in the synthesis of the elastin in the ribosomes. The reactions involved can be written schematically as follows:

The functionality of the cross-link shown would appear to be 8, but this is not the case. The four lysine units involved occur as two pairs, and the two units in each pair are relatively closely spaced along the same chain. The functionality is thus 4, as can be seen from the sketch

An analogous reaction has been carried out on perfluoroelastomers, which are usually very difficult to cross-link. Nitrile side groups placed along the chains are trimerized to triazine, thus giving similarly stable, aromatic cross-links with a functionality of 6. The reaction is

In the case of the elastin, the lysine units involved in the cross-linking are preceded and succeeded by sequences of alanine units. These sequences are thought to be α-helical, which is very intriguing since it suggests that nature carefully positions the potential cross-linking sites spatially as well as sequentially (along the chain backbone).

It should also be mentioned that the uncross-linked precursor protein for elastin is laid down among parallel fibers of crystalline collagen. This gives the cross-linked elastin a fibrous form, but since the elastin is itself not crystalline, these "fibers" are isotropic. Failure to understand this has led some workers to expect highly regular (anisotropic) structures in what is essentially an irregular (isotropic) system.

COMPLICATIONS IN THERMODYNAMICALLY SEMIOPEN SYSTEMS

An *open system* in the thermodynamic sense is one that can exchange its constituent matter with its surroundings (Atkins 1982). *In vivo* elastin (and presumably the other bioelastomers as well) is immersed in excess amounts of aqueous body fluids, which bring it to a state of swelling equilibrium. It is thus a *semiopen system*, which, as mentioned in Chapter 17, can exchange diluent (but not polymer) with its surroundings. As mentioned earlier, changes in temperature or deformation could then significantly change the degree of equilibrium swelling.

Experimental studies have generally been carried out with the elastin swollen with water, to closely approximate *in vivo* conditions. Because of the volatility of water and the open nature of *in vivo* elastin, these measurements have generally been carried out in excess solvent. This has caused a great deal of interpretative difficulty. In one of the most important cases, calorimetric measurements were reported for the elongation of elastin swollen and immersed in excess water (Weis-Fogh and Andersen 1970). Large amounts of heat were given off, and these negative enthalpy changes were held to be inconsistent with the standard random network (described in Chapter 1), in that $|f_e/f|$ was not relatively small. One model proposed to replace it consisted of interconnected protein globules ("oil drops") with water consigned to the spaces between them. Stretching this assembly would distort the spherical droplets to ellipsoids of larger surface area, and the increased exposure of hydrophobic groups to the water would then account for the observed large enthalpy changes (Weis-Fogh and Andersen 1970). Reexamination of the problem, however, showed that the enthalpy changes were dilution effects accompanying the increased swelling that invariably occurs upon elongation of an elastomer in swelling equilibrium with excess solvent (Hoeve and Flory 1974). This conclusion was confirmed by thermoelasticity measurements that were carried out at constant composition ("closed" thermodynamic systems). The experiments utilized either (1) a nonvolatile diluent (Andrady and Mark 1980) or (2) water with the water-swollen samples totally immersed in oil (Gosline 1987). Also relevant are some theoretical calculations based on Monte Carlo simulations, rotational isomeric state theory (Flory 1969), and the standard random network model. They showed that no unusual structures such as globular droplets or β-spirals (Urry et al. 1983) are required to reproduce the experimental thermoelastic results (DeBolt and Mark 1987b).

The basic mechanism for the elasticity of elastin is thus the usual entropy reduction accompanying the deformation of random network chains. Nonetheless, hydrophobic interactions and their change with deformation could have an effect on the elastic free energy of a swollen elastin network. They thus would contribute to the elastic force exhibited by elastin in an open system, such as occurs in *in vivo* deformations (Gosline 1987).

STRESS–STRAIN BEHAVIOR

Stress–strain isotherms obtained on swollen strips of elastin are generally different in shape from those observed for nonbiological elastomers. Specifically, they are not of the form shown in Figure 1.7, frequently even at relatively low elongations. This is apparently due to the fact that such a strip is made up of a multitude of elastin fibers that are generally not "in register" (Hoeve and Flory 1974). That is, different fibers have different amounts of slackness in the undeformed strip and thus experience different elongations as the deformation proceeds. The problem has been solved using single elastin fibers (also swollen), carrying out the measurements microscopically (Aaron and Gosline 1981).

The elongation isotherms thus obtained show some upturns in reduced stress at high elongations. The increases in stress as the rupture point is approached represent increases in ultimate strength, and could therefore be very advantageous to the organism. This increase in strength is presumably a non-Gaussian effect, due to the limited extensibility of the elastin chains. This interpretation is supported by the pattern of theoretical stress–strain isotherms calculated from distribution functions generated using Monte Carlo simulations applied to a rotational isomeric state model of typical elastin sequences (DeBolt and Mark 1988).

The ultimate strength of elastin in body tissues is generally enhanced by crystalline collagen fibers threading through the elastomeric matrix (Gosline 1980). This type of fibrous reinforcement is reminiscent of the webs of nylon fibers embedded in commercial high-pressure tubing to increase resistance to bursting.

STORAGE OF ELASTIC ENERGY

Efficient storage of elastic energy is particularly important in the case of living organisms. Minimizing the amount of energy wastefully degraded into heat during the deformation process maximizes, of course, the organism's chances of survival (Gosline 1980, 1987). It is also obviously advantageous to minimize heat buildup from reciprocating motions such as those occurring in insect flight. It is therefore not surprising to find that the efficiency of elastic energy storage in bioelastomers is quite high.

In the case of insects having extraordinary jumping abilities, such as the flea, survival also depends on the very rapid release of the stored energy. This release has in fact been found to be very short, typically the order of 1 msec.

The ability of bioelastomers such as elastin to store elastic energy efficiently and to use it almost instantaneously for recovery or retraction is remarkable. These properties could be due to the flexibility of the elastin chains (particularly since they are in the highly swollen state) or perhaps to a simpler (less entangled) network topology resulting from the unique cross-linking procedures, which were described above. Elucidation of these points could lead the way to the design of very attractive nonbiological elastomeric materials.

20

FILLED ELASTOMERS

INTRODUCTION

Elastomers, particularly those that cannot undergo strain-induced crystallization, are generally compounded with a reinforcing filler. The two most important examples are the addition of carbon black to natural rubber and to some synthetic elastomers (Boonstra 1979; Rigbi 1980) and silica to silicone rubbers (Warrick et al. 1979). The advantages obtained include improved abrasion resistance, tear strength, and tensile strength. Disadvantages include increases in hysteresis (and thus heat buildup) and compression set (permanent deformation).

The mechanism of the reinforcement obtained is only poorly understood in molecular terms. The network chains certainly adsorb strongly onto the particle surfaces, which would of course give an increase in the effective degree of cross-linking. This is only part of the effect, however (Mullins and Tobin 1965; Meier 1974), but most additional molecular models seem to be highly speculative. Some elucidation might be obtained by incorporating the fillers in a more carefully controlled manner. We describe several such approaches in the remainder of this chapter.

SOME TYPICAL REACTIONS

Hydrolysis of an alkoxysilane such as tetraethoxysilane (TEOS),

$$Si(OEt)_4 + 2H_2O \rightarrow SiO_2 + 4EtOH \tag{20.1}$$

can be used to precipitate very small, well-dispersed particles of silica into a polymeric material (Clarson and Mark 1987; Mark 1988). A variety of substances, generally bases, catalyze this reaction, which occurs quite readily near room temperature. Another example would be the catalyzed hydrolysis of a titanate,

$$Ti(OPr)_4 + 2H_2O \rightarrow TiO_2 + 4PrOH \tag{20.2}$$

in the *in situ* precipitation of titania. Both the silica and titania thus produced give good reinforcement of a variety of elastomers. These are the same chemical reactions used in the sol–gel technology for preparing ceramics. In this case, the process first gives a swollen gel, which is then dried, fired, and densified into the final monolithic piece of ceramic, typically silica.

Details of the techniques used for the *in situ* precipitation of the particles into elastomers are given in Table 20.1.

Table 20.1   Some Techniques for *in Situ* Precipitation

A. *After Cross-Linking*
1. Cross-link the polymer.
2. Swell with TEOS.
3. Hydrolyze TEOS to SiO_2.
4. Dry the sample.

B. *During Cross-Linking*
1. Dissolve TEOS into a reactive polymer.
2. Condense some TEOS with chains for end-linking.
3. Hydrolyze some TEOS to SiO_2.
4. Dry the sample.

C. *Before Cross-Linking*
1. Dissolve TEOS into an inert polymer.
2. Hydrolyze TEOS to SiO_2.
3. Dry the sample.
4. Cross-link the filled polymer.

FILLER PRECIPITATION AFTER CURING

In this technique the polymer is first cured or vulcanized into a network structure using any of the well-known cross-linking techniques described in Chapter 3 (Mark 1988). The network is then swelled with the silane or related molecule to be hydrolzyed, after which it is exposed to water at room temperature, in the presence of a catalyst, for a few hours. The swollen sample can be either placed directly into an excess of water containing the catalyst, or merely exposed to the vapors from the catalyst–water solution. Drying the sample then gives an elastomer filled, and thus reinforced, with the ceramic particles resulting from the hydrolysis reaction.

Although a phase-transfer catalyst can be used in such a reaction, it was found to be unnecessary at least for relatively small specimens. Large samples could, of course, have a nonuniform distribution of particles, a possibility being investigated by solid-state ^{29}Si NMR spectroscopy.

Different alkoxysilanes can swell an elastomeric network to different extents and can hydrolyze at different rates. TEOS seems to be the best for the technique of filler precipitation after curing, as judged by the amount of silica precipitated and the extent of reinforcement obtained. Using the same criteria, basic catalysts seem more effective than acidic ones. Some preliminary studies on the effects of catalyst concentration in particular and the hydrolysis kinetics in general have been carried out. It is found that the rate of particle precipitation can vary in a complex manner, possibly due to the loss of colloidal silica and partial deswelling of the networks when placed into contact with the catalyst solution.

Most of the studies to date have been carried out on poly(dimethylsiloxane) (PDMS) because of the great extent to which its networks swell in TEOS. The same technique has, however, been shown to give good reinforcement of polyisobutylene elastomers. Titanates have been used in the place of silanes, with the resulting titania particles also giving significant improvements in elastomeric properties.

FILLER PRECIPITATION DURING CURING

It is also possible to mix hydroxyl-terminated chains (such as those of PDMS) with excess TEOS, which then serves simultaneously to tetrafunctionally end-link the PDMS into a network structure and to act as the source of silica upon hydrolysis (Mark 1988). This simultaneous curing and filling technique has been

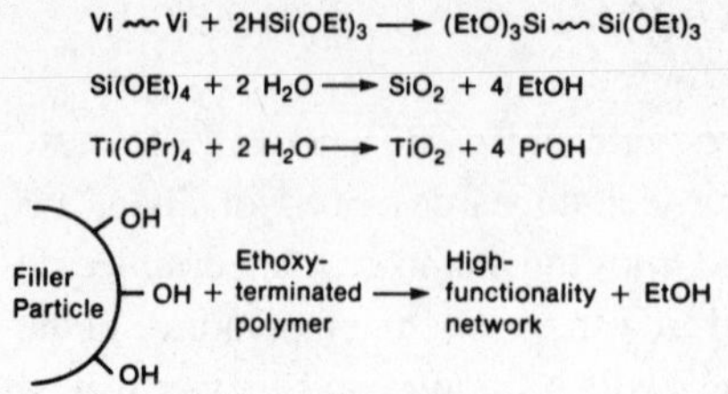

Vi ⏤ Vi + 2HSi(OEt)$_3$ ⟶ (EtO)$_3$Si ⏤ Si(OEt)$_3$

Si(OEt)$_4$ + 2 H$_2$O ⟶ SiO$_2$ + 4 EtOH

Ti(OPr)$_4$ + 2 H$_2$O ⟶ TiO$_2$ + 4 PrOH

Filler Particle ⟨OH, OH, OH⟩ + Ethoxy-terminated polymer ⟶ High-functionality network + EtOH

Figure 20.1 Process for cross-linking (end-linking) triethoxysilyl-terminated poly(dimethylsiloxane) (PDMS) chains using reactive surface groups on silica or titania filler particles (Sur and Mark 1985; Mark 1988).

successfully used for PDMS elastomers having a unimodal distribution of chain lengths and for PDMS elastomers and thermosets having bimodal distributions.

The roles of polymer and alkoxysilane may also be reversed by putting triethoxylsilyl groups at the ends of PDMS chains, as illustrated in Figure 20.1. Reactive groups at the surface of the *in-situ*-generated silica or titania particles then react with the chain ends to simultaneously cure and reinforce the elastomeric material (Sur and Mark 1985).

FILLER PRECIPITATION BEFORE CURING

In the previous two techniques, removal of the unreacted TEOS and the ROH by-product causes a significant decrease in volume, which could be disadvantageous in some applications. This problem can be overcome by precipitating the particles into a polymer that is inert under the hydrolysis conditions, for example, vinyl-terminated PDMS. The resulting polymer–filler suspension, after removal of the other materials, is quite stable. It can be subsequently cross-linked, for example, by silane reaction with the vinyl groups, with only the usual, very small change in volume occurring in any curing process (Mark 1988).

MODIFIED FILLER PARTICLES

If an *in-situ*-filled elastomer is extracted with a good solvent, its modulus and ultimate strength are frequently significantly increased. The effect is probably due to hydrolytic formation of additional reactive groups on the particle surface or to removal of absorbed small molecules, thus increasing the number of sites for particle–polymer bonding.

In some applications, it may be advantageous for the filler particles to have some deformability. It may be possible to induce such deformability by using a molecule that is only partially hydrolyzable, for example, a triethoxysilane [R'Si(OR)$_3$], where R' could be methyl, ethyl, vinyl, or phenyl (Mark 1988).

TYPICAL IMPROVEMENTS FROM *in Situ* PRECIPITATIONS

Some typical results in the Mooney–Rivlin representation are shown in Figure 20.2 (Ning and Mark 1984). Generating the filler particles *in situ* greatly increases the elastomer modulus and introduces an upturn in it prior to rupture. Also, as mentioned in the preceding section, extraction with a solvent gives further significant improvements.

Another typical representation shows the nominal stress as a function of elongation, as illustrated for titania-filled PDMS in Figure 20.3 (Wang and

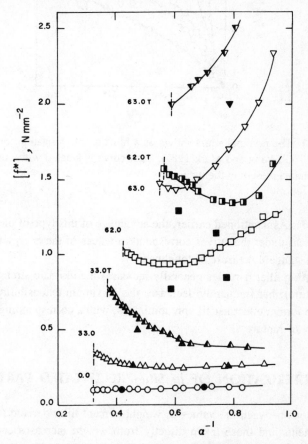

Figure 20.2 Modulus shown as a function of reciprocal elongation for unfilled and filled PDMS networks at 25°C (Ning and Mark 1984). The numbers correspond to the percent by weight of filler in the network, and the letter T specifies treatment (extraction) with tetrahydrofuran. Filled symbols are for results obtained out of sequence to test for reversibility, and the vertical dashed lines locate the rupture points.

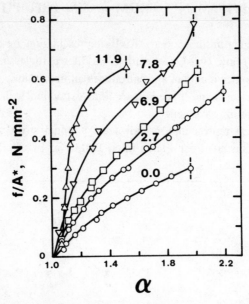

Figure 20.3 The nominal stress shown as a function of elongation for PDMS networks at 25°C (Wang and Mark 1987). Each curve is labeled with the percent by weight of titania present in the network.

Mark 1987). As mentioned earlier, the advantage of this type of plot is the fact that the areas under the curves correspond to values of the energy required for rupture, a standard measure of toughness.

Generating filler particles generally increases the ultimate strength ($[f^*]$ or f^* at rupture), but frequently decreases the maximum extensibility (α at rupture). The former effect usually predominates, with a corresponding increase in the energy of rupture.

CHARACTERIZATION OF *in Situ*–PRECIPITATED PARTICLES

Comparisons between the values of weight-percent filler derived from density measurements and those taken directly from weight increases can give very useful information on the filler particles. For example, the fact that the former estimate is smaller than the latter in the case of silica-filled PDMS elastomers indicates that there are probably either voids or unreacted organic groups in the filler particles.

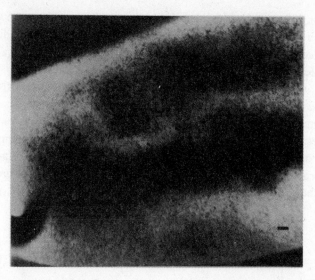

Figure 20.4 Transmission electron micrograph of a PDMS network containing well-defined silica particles obtained in an ethylamine (base) catalyzed hydrolysis of tetraethoxylsilane (TEOS) (Mark et al. 1985). The length of the bar in the lower right-hand corner of this and the following figure corresponds to 1000 Å.

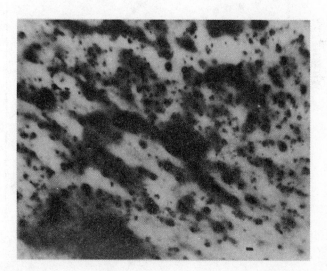

Figure 20.5 Electron micrograph of a PDMS network containing "fuzzy" silica particles obtained in an acetic acid catalyzed hydrolysis of TEOS (Mark et al. 1985).

151

The transmission electron micrograph shown in Figure 20.4 shows that the particles in this silica-filled PDMS network have (1) an average diameter of approximately 80 Å (a very desirable size for reinforcement), (2) a relatively narrow size distribution, (3) very little of the agglomeration that is usually a problem is filler-blended elastomers, and (4) well-defined surfaces (Mark et al. 1985). The good definition generally occurs when the catalyst is a base, as was the ethylamine used for this sample. Use of an acidic catalyst, on the other hand, gives poorly defined, fuzzy-looking particles, as illustrated in Figure 20.5. This lack of definition is consistent with results in the sol–gel–ceramics area, where researchers have concluded that acidic catalysts give structures that are less branched and less compact than those obtained from basic catalysts.

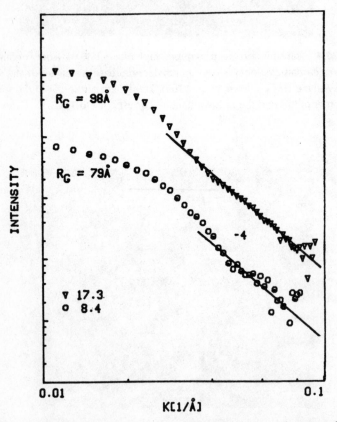

Figure 20.6 Intensity of small-angle X-ray scattering (SAXS) as a function of the scattering vector for PDMS networks containing 17.3 (▽) and 8.4 (○) wt% silica (Mark 1988). The labels give the values for the radius of gyration R_G and the terminal slope.

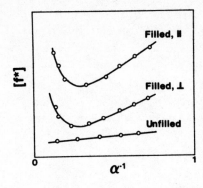

Figure 20.7 Schematic stress–strain iso-therms showing reinforcement obtained from magnetic filler particles parallel (‖) and per-pendicular (⊥) to the lines of force of a mag-netic field present during the cross-linking process (Rigbi and Mark 1985; Sohoni and Mark 1987).

Some typical small-angle X-ray scattering (SAXS) results are shown in Fig-ure 20.6 (Mark 1988). The radii of gyration thus obtained can be correlated, for example, with electron microscopy results and with various elastomeric properties. Also, the shapes of the curves can give information on the distri-bution of particle sizes, and the terminal slopes can indicate whether the parti-cles are well defined (slope of -4) or poorly defined (slope of -3). Experi-ments using neutron scattering should also be very useful in characterizing ceramic particles of this type.

MAGNETIC FILLER PARTICLES

Some filler particles can be manipulated with a magnetic field. For example, magnetic ferrite particles dispersed in PDMS can be aligned in a magnetic field during the curing process. In this way, anisotropic mechanical properties can be obtained, even from essentially spherical particles (Rigbi and Mark 1985; Sohoni and Mark 1987). Some typical results are shown schematically in Figure 20.7. The reinforcement is seen to be significantly larger in the direction par-allel to the magnetic lines of force.

This technique can be combined with the *in situ* approach by generating the magnetic particles in the presence of a magnetic field, for example, by ther-molysis or photolysis of a metal carbonyl.

21

CURRENT PROBLEMS AND NEW DIRECTIONS

Table 21.1 lists some molecular aspects of rubberlike elasticity that clearly need additional research (Mark 1984a).

One of the necessary (but not sufficient) conditions for rubberlike elasticity, mentioned in Chapter 1, is that the polymer be above its glass transition temperature. Also, if the polymer would otherwise be highly crystalline, it should obviously be above its melting point. Since these two transition temperatures, T_g and T_m, in part delimit the temperature range for this type of elasticity, an improved molecular understanding of their dependence on structure could help enormously in designing new elastomeric materials.

High-performance elastomers are those that remain elastomeric at very low temperatures and are relatively stable at very high temperatures. Some phosphazene polymers, shown schematically in Figure 21.1 (Allcock 1977; Mark and Yu 1977; Hsu and Mark 1987), are in this category. These polymers have rather low glass transition temperatures, which is unusual since the skeletal bonds of the chains are thought to have some double-bond character. There are thus a number of interesting problems related to the configurational characteristics and elastomeric behavior of these semi-inorganic polymers.

The development and exploitation of new cross-linking approaches could be of considerable importance. Some of the techniques described in Chapter 3, such as end-linking of chains and aggregation of parts of triblock copolymers,

Table 21.1 Some Important Areas for Investigation

1. Improved understanding of dependence of T_g and T_m on polymer structure
2. Preparation and characterization of high-performance elastomers
3. New cross-linking techniques
4. Improved understanding of network topology
5. More experimental results for deformations other than elongation and swelling
6. Better characterization of segmental orientation
7. More detailed understanding of critical phenomena and gel collapse
8. Additional molecular characterization using NMR spectroscopy and various scattering techniques
9. Study of possibly novel properties of bioelastomers
10. Improved molecular understanding of reinforcing effects of filler particles in elastomers

are steps in the right direction, but we are far from the sophistication shown by nature in the preparation of bioelastomers.

As mentioned in Chapters 5, 10, and 11, one of the most important unsolved problems in rubberlike elasticity concerns network topology. Specifically, what are the roles of entanglements, both in the vicinities of the cross-links and relatively far from the cross-links along the chain trajectories?

The molecular theories can predict elastic equations of state for any type of deformation whatever, as illustrated in Chapter 6 and Appendix C. Yet almost all experimental studies have involved either elongation or swelling equilibrium, presumably because of the ease with which these deformations can be imposed and characterized. For deformations other than elongation and swelling, there is a real need for reliable experimental results that will allow more extensive comparisons between theory and experiment.

There is also a need for better characterization of segmental orientation by distinguishing intermolecular orientation correlations from those that are intramolecular. The former show their presence through an enhancement of orientation, but such local correlations do not affect the stress (Queslel et al. 1985). A more basic understanding of intermolecular contributions to local segmental orientations is obviously required.

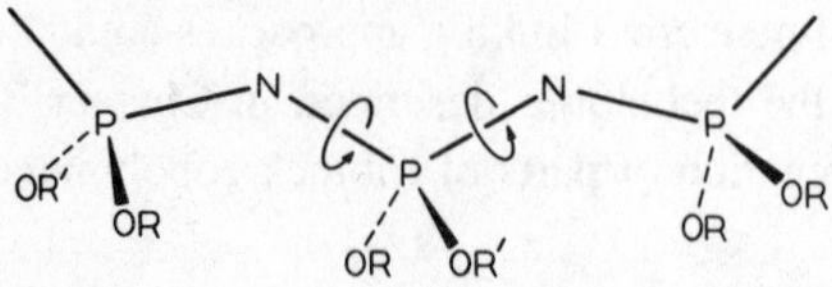

Figure 21.1 Structure of a semi-inorganic phosphazene polymer (Mark and Yu 1977), where R represents either alkyl or aryl groups.

Critical phenomena and gel collapse lend themselves to modern theoretical treatments such as renormalization group theory and scaling arguments. At the same time, they play a basic role in many biological events. An improved understanding, both theoretical and experimental, would be highly desirable.

Nuclear magnetic resonance (NMR) spectroscopy and small-angle neutron scattering (SANS) are two advanced experimental techniques now available for the characterization of network behavior. These techniques give information at the molecular level, in contrast to the macroscopic information provided by most other characterization techniques. They thus should be exploited more in attempting to obtain a better molecular understanding of rubberlike elasticity.

As we described in Chapter 19, bioelastomers can have intriguing structures and novel mechanical properties. A better understanding of the corresponding structure–property relationships could be as important as it would be interesting.

A particularly challenging problem is the development of a more quantitative molecular understanding of the effects of filler particles, such as carbon black in natural rubber and silica in silicone polymers. As described in Chapter 20, such fillers provide tremendous reinforcement in elastomers; how fillers do this is still only poorly comprehended. It is difficult to imagine another problem of comparable fundamental and practical significance in the area of rubberlike elasticity.

APPENDIXES

APPENDIX A

RELATIONSHIPS BETWEEN ν, ξ, AND M_c

In this section we derive Eqs. (3.2) and (3.3) for a perfect network and then discuss their forms for an imperfect network.

An acyclic giant molecule or a tree (Flory 1976) forms a convenient starting point. An example of such a tree is shown in Figure A.1. Lines denote chains, and dots represent labeled points that join with each other to form junctions or cross-links.

Formation of ξ connections of labeled points reduces their number to $\nu - \xi + 1 \simeq \nu - \xi$ and also introduces ξ independent cyclic paths. The number of such cyclic paths, ξ, is the *cycle rank* of the network, and ν is the number of network chains. In a perfect network there are no dangling chains and loops, and the functionality of each junction is greater than 2. If the labeled points of a tree are connected to form a perfect network, then the number μ of junctions is given by

$$\mu = \nu - \xi + 1 \simeq \nu - \xi \qquad (A.1)$$

The reader may verify Eq. (A.1) easily by forming a sample perfect network from a tree. Eliminating μ from Eqs. (A.1) and (3.1) leads to Eq. (3.2). An alternate form of this equation,

$$\xi = (\phi/2 - 1)\mu \qquad (A.2)$$

is obtained by eliminating ν between Eqs. (A.1) and (3.1).

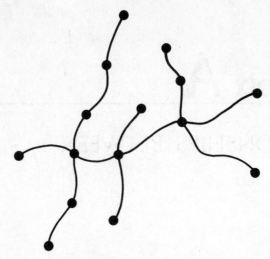

Figure A.1 A giant acyclic molecule or a tree. Lines denote chains, and dots are labeled points capable of reacting with other labeled points to form junctions.

The density ρ of the network is defined as

$$\rho = \nu M_c / N_A V_0 \tag{A.3}$$

where M_c is the average molecular weight between junctions, N_A is Avogadro's number, and V_0 is the volume of the perfect network during its formation. If the network is formed in solution, V_0 is the total volume of polymer and solvent, and ρ is the density in this state. Substituting ν from Eq. (A.3) into Eq. (3.2) leads to Eq. (3.3).

The cycle rank ξ is the essential and universal parameter that characterizes the structure and elasticity of a network. The Helmholtz elastic free energy and the force of a network are both proportional to ξ (Duiser and Staverman 1965; Graessley 1975a,b). This is true for an imperfect network as well as a perfect one.

The characterization of the number of chains and junctions that contribute to the elastic activity of an imperfect network is outlined briefly by Flory (1982) and the references cited therein.

Equations (3.1) and (3.2) continue to hold exactly for an imperfect network if ν, μ, and ϕ are redefined. Thus

$$\mu' = 2\nu'/\bar{\phi}$$
$$\xi - 1 \simeq \xi = (1 - 2/\bar{\phi})\nu' \tag{A.4}$$

where μ' is the number of junctions of functionality 3 or more, and ν' is the total number of chains excluding loops and dangling chains. Two chains separated by one bifunctional junction or by a series of them are considered to be a single chain and therefore contribute only unity to ν'. $\overline{\phi}$ is the average junction functionality. The reader may verify Eqs. (A.4) by constructing an imperfect network.

A theoretical estimate of M_c for all types of imperfections is not definitely established. A simple relation may be written, however, for a tetrafunctionally cross-linked imperfect network with no defects other than chain ends. The cycle rank for such a network is given (Queslel and Mark 1985b) as

$$\xi = \frac{\rho}{2M_c} \left(1 - 3\frac{M_c}{M_n} \right) \tag{A.5}$$

where M_n is the molecular weight of the primary chains before cross-linking. Derivation of Eq. (A.5) requires some discussion, which may be found in the paper by Queslel and Mark (1985b).

APPENDIX B

RELATIONSHIPS BETWEEN $\langle \bar{r}^2 \rangle$, $\langle (\Delta r)^2 \rangle$, $\langle r^2 \rangle_0$, AND ϕ

The end-to-end vector $\mathbf{r}$ of a network chain is expressed in terms of its average value $\bar{\mathbf{r}}$ and instantaneous fluctuations $\Delta \mathbf{r}$ from this mean by Eq. (4.12). According to the phantom network theory (Flory 1976), the instantaneous distributions of $\mathbf{r}$, $\bar{\mathbf{r}}$ and $\Delta \mathbf{r}$, which are assumed to be Gaussian, are

$$W(\mathbf{r}) = (\gamma/\pi)^{3/2} \exp(-\gamma r^2)$$

$$X(\bar{\mathbf{r}}) = (\chi/\pi)^{3/2} \exp(-\chi \bar{r}^2)$$

$$\Psi(\Delta \mathbf{r}) = (\psi/\pi)^{3/2} \exp[-\psi(\Delta r)^2] \tag{B.1}$$

where

$$\gamma = \frac{3}{2\langle r^2 \rangle_0}, \qquad \chi = \frac{3}{2\langle \bar{r}^2 \rangle}, \qquad \psi = \frac{3}{2\langle (\Delta r)^2 \rangle} \tag{B.2}$$

The distribution of $\mathbf{r}$ may be written in terms of the distribution of $\bar{\mathbf{r}}$ and $\Delta \mathbf{r}$ as

$$W(\mathbf{r}) = \int X(\bar{\mathbf{r}}) \, \Psi(\mathbf{r} - \bar{\mathbf{r}}) \, d\bar{\mathbf{r}} \tag{B.3}$$

where the argument of Ψ follows from Eq. (4.12). Replacing $(\Delta r)^2$ in Eq. (B.1) by the square of the magnitude of the vector $(\mathbf{r} - \bar{\mathbf{r}})$, substituting Eqs. (B.1) into Eq. B.3, and performing the integration over $\bar{\mathbf{r}}$ results in

$$1/\psi + 1/\chi = 1/\gamma \tag{B.4}$$

or

$$\langle \bar{r}^2 \rangle + \langle (\Delta r)^2 \rangle = \langle r^2 \rangle_0 \tag{B.5}$$

The parameters ψ and γ are related for a phantom network (Eichinger 1972; Graessley 1975b; Pearson 1977) by the following expression:

$$\psi = (\phi/2)\gamma = 3\phi/4 \langle r^2 \rangle_0 \tag{B.6}$$

Derivation of Eq. (B.6) requires a knowledge of network topology. Replacing ψ in Eq. (B.6) by $3/2 \langle (\Delta r)^2 \rangle$ leads to Eq. 4.15b. Use of Eq. (B.4) or (B.5) together with (B.6) leads to Eq. 4.15a.

APPENDIX C

EQUATIONS OF STATE FOR MISCELLANEOUS DEFORMATIONS FROM THE CONSTRAINED JUNCTION THEORY

The stress in a deformed network is given by Eq. (6.7) in which ΔA_{el} is the elastic free energy of the network. From Chapters 4 and 5 the elastic free energy for the constrained junction model is

$$\Delta A_{\text{el}} = \Delta A_{\text{ph}} + \Delta A_{\text{c}}$$

$$= \frac{\xi kT}{2} \left\{ \sum_t \lambda_t^2 - 3 + \frac{\mu}{\xi} \sum_t [B_t + D_t - \ln (1 + B_t)(1 + D_t)] \right. \quad \text{(C.1)}$$

where

$$B_t = \kappa^2 (\lambda_t^2 - 1)(\lambda_t^2 + \kappa)^{-2}$$

$$D_t = \lambda_t^2 \kappa^{-1} B_t \qquad \text{(C.2)}$$

Substitution into Eq. (6.7) leads to

$$\tau_t = \left(\frac{\xi kT}{2V} \right) \lambda_t \sum_i \left(1 + \frac{\mu}{\xi} K_i \right) \frac{\partial \lambda_i^2}{\partial \lambda_t} \qquad \text{(C.3)}$$

where

$$K_i \equiv K(\lambda_i^2) = B_i[\dot{B}_i(B_i + 1)^{-1} + \kappa^{-1}(\lambda_i^2\dot{B}_i + B_i)(B_i + \kappa\lambda_i^{-2})^{-1}]$$
$$\dot{B}_i \equiv \partial B_i/\partial\lambda_i^2 = B_i[(\lambda_i^2 - 1)^{-1} - 2(\lambda_i^2 + \kappa)^{-1}]$$ (C.4)

For uniaxial extension, substituting Eqs. (6.9a,b) into Eq. (C.3) leads to

$$\tau_1 = \left(\frac{\xi kT}{V}\right)\left(\frac{V}{V_0}\right)^{2/3}\left[\alpha^2 - \alpha^{-1} + \left(\frac{\mu}{\xi}\right)(\alpha^2 K_1 - \alpha^{-1}K_2)\right]$$ (C.5)

The final term in the square brackets in Eq. (C.5) is the term due to constraints on the junction. Without this term, Eq. (C.5) reduces to the phantom network expression given by Eq. (6.11) (with $\mathcal{F} = \xi/2$). The force f acting on the network is obtained by multiplying both sides of Eq. (C.5) by the deformed area. The reduced force then follows from the definition given by Eq. (6.13), as

$$[f^*] = \left(\frac{\xi kT}{V_d}\right)v_{2C}^{2/3}\left[1 + \left(\frac{\mu}{\xi}\right)\frac{\alpha K_1 - \alpha^{-2}K_2}{\alpha - \alpha^{-2}}\right]$$ (C.6)

For biaxial extension, using Eqs. (6.17) in Eq. (C.3) leads to

$$\tau_1 = \left(\frac{\xi kT}{V}\right)\left(\frac{V}{V_0}\right)^{2/3}\left[\alpha_1^2 - \frac{1}{\alpha_1^2\alpha_2^2} + \left(\frac{\mu}{\xi}\right)\left(\alpha_1^2 K_1 - \frac{K_3}{\alpha_1^2\alpha_2^2}\right)\right]$$
$$\tau_2 = \left(\frac{\xi kT}{V}\right)\left(\frac{V}{V_0}\right)^{2/3}\left[\alpha_2^2 - \frac{1}{\alpha_1^2\alpha_2^2} + \left(\frac{\mu}{\xi}\right)\left(\alpha_2^2 K_2 - \frac{K_3}{\alpha_1^2\alpha_2^2}\right)\right]$$ (C.7)

The stresses in pure shear are obtained by letting $\alpha_2 = 1$ and $\alpha_1 = \alpha$ in Eq. (C.7):

$$\tau_1 = \left(\frac{\xi kT}{V}\right)\left(\frac{V}{V_0}\right)^{2/3}\left[\alpha^2 - \frac{1}{\alpha^2} + \left(\frac{\mu}{\xi}\right)\left(\alpha^2 K_1 - \frac{K_3}{\alpha^2}\right)\right]$$
$$\tau_2 = \left(\frac{\xi kT}{V}\right)\left(\frac{V}{V_0}\right)^{2/3}\left[1 - \frac{1}{\alpha^2} + \left(\frac{\mu}{\xi}\right)\left(K_2 - \frac{K_3}{\alpha^2}\right)\right]$$ (C.8)

APPENDIX D

FORTRAN PROGRAM FOR CALCULATING [f*]

The following Fortran program calculates the reduced stress [f*] from the constrained junction theory for a network in simple elongation. The output is values of [f*] as a function of α for a network characterized by ϕ, v_2, v_{2C}, $\xi kT/V_0$, and κ. Calculations are made according to Eq. (C.6).

```
C  ***  CALCULATION OF REDUCED STRESS AS A
C  ***  FUNCTION OF PHI, V2, V2C, MODS, KAPPA, AND ALPHA
C
C  ***  PHI = JUNCTION FUNCTIONALITY
C  ***  V2 = FRACTION OF POLYMER DURING STRETCHING
C  ***  V2C = FRACTION OF POLYMER DURING CROSS-LINKING
C  ***  MODS = PHANTOM MODULUS (XI*K*T/V0)
C  ***  KAPPA = KAPPA PARAMETER
C  ***  ALPHA = EXTENSION RATIO (FINAL LENGTH/INITIAL SWOLLEN
        LENGTH)
C
        PROGRAM FSTAR
        REAL FSTAR, ALINV
        WRITE(*,*) 'INPUT THE DATA IN THE FOLLOWING ORDER'
        WRITE(*,*)
        WRITE(*,*) 'PHI, V2, V2C, ZMOD, KAPPA, ALLO, ALUP, ALINC'
        READ(*,*) PHI, V2, V2C, ZMOD, PKAPPA, ALLO, ALUP, ALINC
```

```
C
C  *** ALLO = LOWER LIMIT OF ALPHA
C  *** ALUP = UPPER LIMIT OF ALPHA
C  *** ALINC = INCREMENTS OF ALPHA
C
       AL = ALLO
  100  ZS = (V2C/V2)**(2./3.)*AL**2
       CALL GETK (PKAPPA, ZS, FK1)
       ZS = (V2C/V2)**(2./3.)/AL
       CALL GETK (PKAPPA, ZS, FK2)
C
       F1 = 2./(PHI − 2.)
       F2 = AL * FK1 − FK2/AL**2
       F3 = 1./(AL − 1./AL**2)
       FSTAR = ZMOD * (1. + F1*F2*F3)
       ALINV = 1./AL
C
       WRITE(*,*) AL, ALINV, FSTAR
C
       AL = AL + ALINC
       IF (AL.LE.ALUP) GO TO 100
       STOP
       END
C
       SUBROUTINE GETK (SKAPPA, ZS, ZZ)
C
       B = SKAPPA**2*(ZS − 1.)/(ZS + SKAPPA)**2
       D = ZS*B/SKAPPA
       BD = B*(1./(ZS − 1.) − 2./(ZS + SKAPPA))
C
       DD = (ZS*BD + B)/SKAPPA
       ZZ = B*BD/(1. + B) + D*DD/(1. + D)
C
       RETURN
       END
```

APPENDIX E

ILLUSTRATIVE CALCULATIONS OF M_c

(The discussion and calculations for this appendix are taken from Mark (1982b).)

ANALYSIS OF TYPICAL ELONGATION OR COMPRESSION DATA

Experimental data in simple tension or compression are usually presented in terms of the familiar Mooney–Rivlin plot, where the reduced force is plotted against reciprocal extension ratio as shown in Figure 8.1. If the upturn in this Figure (due to crystallization or finite chain extensibility) is ignored, the curve may be extrapolated to $\alpha^{-1} = 0$ to obtain an approximation of the reduced stress for the phantom network, represented by Eq. (6.14) with $\mathcal{F} = \xi/2$, as

$$[f^*]_{ph} = \left(\frac{\xi}{V_d}\right) kTv_{2C}^{2/3} \tag{E.1}$$

Thus $[f^*]_{ph}$ represents the shear modulus of the phantom network. It may be expressed in terms of chain density by substituting Eq. (3.2) into Eq. (E.1):

$$[f^*]_{ph} = \left(1 - \frac{2}{\phi}\right) \frac{\nu}{V_d} kTv_{2C}^{2/3} \tag{E.2}$$

171

Substituting Eq. (A.2) into Eq. (E.1) relates $[f^*]_{ph}$ to the junction density,

$$[f^*]_{ph} = \left(\frac{\phi}{2} - 1\right) \frac{\mu}{V_d} kTv_{2C}^{2/3} \tag{E.3}$$

and the use of Eq. (3.3) in Eq. (E.1) leads to an expression containing the molecular weight M_c of the network chains

$$[f^*]_{ph} = \left(1 - \frac{2}{\phi}\right) \frac{\rho RT}{M_c} \tag{E.4}$$

where ρ is the density of the bulk polymer.

Experimental determination of $[f^*]_{ph}$ from the Mooney–Rivlin plot thus allows one to derive values for v, μ, and M_c according to Eqs. (E.2), (E.3) and (E.4).

Suppose a network having tetrafunctional cross-links ($\phi = 4$) that were introduced in the undiluted state ($v_{2S} = 1.00$) has a reduced force of

$$[f^*]_{ph} = 0.10 \text{ N mm}^{-2}$$

at 298.2 K. [1.00 N mm^{-2} $= 10^6$ N m^{-2} (or Pa) $= 1$ MN m^{-2} (or MPa) $= 10.2$ kg cm^{-2}.] Taking k as 1.381×10^{-20} N mm K^{-1} chain^{-1} and using the above data in Eq. (E.2) leads to

$$v/V = 4.86 \times 10^{16} \text{ chains mm}^{-3}$$

Use of Avogadro's number $N_A = 6.02 \times 10^{23}$ mol^{-1} then gives

$$v/V = 8.06 \times 10^{-8} \text{ mols of chains mm}^{-3}$$

From Eq. (E.3), we obtain

$$\mu/V = 4.03 \times 10^{-8} \text{ mols of cross-links mm}^{-3}.$$

If the polymer has a density $\rho = 0.900$ g cm^{-3} (or 9.00×10^{-4} g mm^{-3}), then Eq. (E.4) gives

$$M_c = 1.12 \times 10^4 \text{ g mol}^{-1}$$

ANALYSIS OF TYPICAL SWELLING DATA

A typical network studied in this regard might also have been tetrafunctionally cross-linked in the undiluted state, have the same value of ρ, and exhibit an equilibrium degree of swelling characterized by $v_{2m} = 0.100$ in a solvent having a molar volume

$$V_1 = 80 \text{ cm}^3 \text{ mol}^{-1} \text{ (or } 8.00 \times 10^4 \text{ mm}^3 \text{ mol}^{-1})$$

and an interaction parameter with the polymer corresponding to

$$\chi = 0.30$$

Substituting these data into Eq. (7.10) leads to a value for the molecular weight of the network chains:

$$M_c = 0.708 \times 10^4 \text{ g mol}^{-1}$$

Using the relationships between M_c, μ, and ν and the above data gives

$$\left(\frac{\mu}{V}\right) = \frac{2}{\phi}\frac{\rho}{M_c} = 6.36 \times 10^{-8} \text{ mols of cross-links mm}^{-3}$$

and

$$(\nu/V) = \rho/M_c = 12.7 \times 10^{-8} \text{ mols of chains mm}^{-3}.$$

Results calculated using the more complicated constrained junction equation with a reasonable value of κ are not very different from those calculated from Eq. (7.10).

REFERENCES

Aaron, B. B. and Gosline, J. M. (1981) *Biopolymers*, **20,** 1247.

Abe, Y. and Flory, P. J. (1970) *J. Chem. Phys.*, **52,** 2814.

Abe, Y. and Flory, P. J. (1971a) *Macromolecules*, **4,** 219.

Abe, Y. and Flory, P. J. (1971b) *Macromolecules*, **4,** 230.

Aklonis, J. J. and MacKnight, W. J. (1983) *Introduction to Polymer Viscoelasticity*, 2nd ed., Wiley-Interscience, New York.

Allcock, H. R. (1977) *Angew. Chem., Int. Ed. Eng.*, **16,** 147.

Allegra, G. (1980) *Makromol. Chem.*, **181,** 1127.

Allen, G., Kirkham, M. J., Padget, J., and Price, C. (1971) *Trans. Faraday Soc.*, **67,** 1278.

Andrady, A.L. and Mark, J. E. (1980) *Biopolymers*, **19,** 849.

Andrady, A. L., Llorente, M. A., and Mark, J. E. (1980) *J. Chem. Phys.*, **72,** 2282.

Andrady, A. L., Llorente, M. A., Sharaf, M. A., Rahalkar, R. R., Mark, J. E., Sullivan, J. L., Yu, C. U., and Falender, J. R. (1981) *J. Appl. Polym. Sci.*, **26,** 1829.

Atkins, P. W. (1982) *Physical Chemistry*, Freeman, San Francisco.

Bagrodia, S., Tant, M. R., Wilkes, G. L., and Kennedy, J. P. (1987) *Polymer*, **28,** 2207.

Bahar, I. and Erman, B. (1987) *Macromolecules*, **20,** 1696.

Bahar, I., Erbil, Y., Baysal, B., and Erman, B. (1987) *Macromolecules*, **20,** 1353.

Ball, R. C., Doi, M., and Edwards, S. F. (1981) *Polymer*, **22,** 1010.

Bastide, J., Duplessix, R., Picot, C., and Candau, S. (1984) *Macromolecules*, **17,** 83.

Beltzung, M., Picot, C., Rempp, P., and Herz, J. (1982) *Macromolecules*, **15,** 1594.

Beltzung, M., Picot, C., and Herz, J. (1984) *Macromolecules*, **17,** 663.

Bennett, R. L., Keller, A., and Stejny, J. (1976) *J. Polym. Sci., Polym. Chem. Ed.*, **14,** 3021, 3027.

Benoit, H., Decker, D., Duplessix, R., Picot, C., Rempp, P., Cotton, J. P., Farnoux, B., Jannink, G., and Ober, R. (1976) *J. Polym. Sci., Polym. Phys. Ed.*, **14,** 2119.

Beshah, K., Mark, J. E., Himstedt, A., and Ackerman, J. L. (1986) *J. Polym. Sci., Polym. Phys. Ed.*, **24,** 1207.

Birshtein, T. M. and Ptitsyn, O. B. (1966) *Configurations of Macromolecules*, Wiley-Interscience, New York.

Boonstra, B. B. (1979) *Polymer*, **20,** 691.

Borg, E. L. (1973) In *Rubber Technology*, M. Morton, Ed., 2nd ed., Van Nostrand Reinhold, New York.

Boué, F. and Vilgis, Th. (1986) *Colloid Polym. Sci.*, **264,** 285.

Brotzman, R. W. and Flory, P. J. (1987) *Macromolecules*, **20,** 351.

Brydson, J. A. (1978) *Rubber Chemistry*, Applied Science Publishers, London.

Bueche, F. (1962) *Physical Properties of Polymers*, Wiley-Interscience, New York.

Candau, S., Bastide, J., and Delsanti, M. (1982) *Adv. Polym. Sci.*, **44,** 27.

Chiu, D. S. and Mark, J. E. (1977) *Colloid Polym. Sci.*, **254,** 644.

Clarson, S. J. and Mark, J. E. (1987) *Polym. Commun.*, **28,** 249.

Clarson, S. J., Mark, J. E., and Semlyen, J. A. (1986) *Polym. Commun.*, **27,** 244.

Clough, S. B., Maconnachie, A., and Allen, G. (1980) *Macromolecules*, **13,** 774.

Coran, A. Y. (1978) In *Science and Technology of Rubber*, F. R. Eirich, Ed., Academic Press, New York.

Curro, J. G. and Mark, J. E. (1984) *J. Chem. Phys.*, **80,** 4521.

Davis, W. D. (1973) In *Rubber Technology*, M. Morton, Ed., 2nd ed., Van Nostrand Reinhold, New York.

DeBolt, L. C. and Mark, J. E. (1987a) *Macromolecules*, **20,** 2369.

DeBolt, L. C. and Mark, J. E. (1987b) *Polymer*, **28,** 416.

DeBolt, L. C. and Mark, J. E. (1988) *J. Polym. Sci., Polym. Phys. Ed.*, **26,** 865.

de Gennes, P.-G. (1979) *Scaling Concepts in Polymer Physics*, Cornell University Press, Ithaca, New York.

Deloche, B., and Samulski, E. T. (1981) *Macromolecules*, **14,** 575.

Dole, M., Ed. (1972) *The Radiation Chemistry of Macromolecules*, Vols. 1 and 2, Academic Press, New York.

Dondos, A. and Benoit, H. (1971) *Macromolecules*, **4,** 279.

Duiser, J. A. and Staverman, A. J. (1965) In *Physics of Noncrystalline Solids*, J. A. Prins, Ed., North Holland Publishing, Amsterdam.

Dušek, K. (1986) *Adv. Polym. Sci.*, **78**, 1.

Eichinger, B. E. (1972) *Macromolecules*, **5**, 496.

Eichinger, B. E. (1983) *Annu. Rev. Phys. Chem.*, **34**, 359.

Eisenberg, A. and King, M. (1977) *Ion-Containing Polymers*, Academic Press, New York.

Elias, H.-G. (1977) *Macromolecules*, Vol. 2, Plenum Press, New York.

Erman, B. (1981) *J. Polym. Sci., Polym. Phys. Ed.*, **19**, 829.

Erman, B. (1987) *Macromolecules*, **20**, 1917.

Erman, B. and Flory, P. J. (1978) *J. Polym. Sci., Polym. Phys. Ed.*, **16**, 1115.

Erman, B. and Flory, P. J. (1982) *Macromolecules*, **15**, 806.

Erman, B. and Flory, P. J. (1983) *Macromolecules*, **16**, 1601, 1607.

Erman, B. and Flory, P. J. (1986) *Macromolecules*, **19**, 2342.

Erman, B. and Mark, J. E. (1987) *Macromolecules*, **20**, 2892.

Erman, B. and Monnerie, L. (1985) *Macromolecules*, **18**, 1985.

Ferry, J. D. (1980) *Viscoelastic Properties of Polymers*, 3rd ed., Wiley, New York.

Flory, P. J. (1942) *J. Chem. Phys.*, **10**, 51.

Flory, P. J. (1947) *J. Chem. Phys.*, **15**, 397.

Flory, P. J. (1953) *Principles of Polymer Chemistry*, Cornell University Press, Ithaca, New York.

Flory, P. J. (1961) *Trans. Faraday Soc.*, **57**, 829.

Flory, P. J. (1969) *Statistical Mechanics of Chain Molecules*, Wiley-Interscience, New York.

Flory, P. J. (1973) *Pure Appl. Chem., Macro. Chem.-8*, **33**, 1.

Flory, P. J. (1976) *Proc. R. Soc. London, A*, **351**, 351.

Flory, P. J. (1977a) *J. Chem. Phys.*, **66**, 5720.

Flory, P. J. (1977b) In *Contemporary Topics in Polymer Science*, Vol. 2, E. M. Pearce, and J. R. Schaefgen, Eds., Plenum, New York.

Flory, P. J. (1979) *Polymer*, **20**, 1317.

Flory, P. J. (1982) *Macromolecules*, **15**, 99.

Flory, P. J. (1984) *Pure Appl. Chem.*, **56**, 305.

Flory, P. J. (1985a) *Polym. J. (Tokyo)*, **17**, 1.

Flory, P. J. (1985b) *Brit. Polym. J.*, **17**, 96.

Flory, P. J. and Chang, V. W. C. (1976) *Macromolecules*, **9**, 33.

Flory, P. J. and Erman, B. (1982) *Macromolecules*, **15**, 800.

Flory, P. J. and Rehner, J., Jr. (1943) *J. Chem. Phys.*, **11**, 512, 521.

Flory, P. J., Ciferri, A., and Hoeve, C. A. J. (1960) *J. Polym. Sci.*, **45**, 235.

Galiatsatos, V. and Mark, J. E. (1987) *Macromolecules*, **20**, 2631.

Garrido, L. and Mark, J. E.(1985) *J. Polym. Sci., Polym. Phys. Ed.*, **23**, 1933.

Garrido, L., Mark, J. E., Clarson, S. J., and Semlyen, A. J. (1985a) *Polym. Commun.*, **26**, 53.

Garrido, L., Mark, J. E., Clarson, S. J., and Semlyen, A. J. (1985b) *Polym. Commun.*, **26**, 55.

Gaylord, R. J. (1976) *J. Polym. Sci., Polym. Phys. Ed.*, **14**, 1827.

Gaylord, R. J. (1982) *Polym. Bulletin*, **8**, 325.

Gee, G. (1980) *Macromolecules*, **13**, 705.

Gee, G., Stern, J., and Treloar, L. R. G. (1950) *Trans. Faraday Soc.*, **46**, 1101.

Gent, A. N. (1969) *Macromolecules*, **2**, 262.

Godovsky, Yu, K. (1986) *Adv. Polym. Sci.*, **76**, 31.

Gosline, J. M. (1980) In *The Mechanical Properties of Biological Materials* (Symposium XXXIV), J. F. V. Vincent, and J. D. Currey, Eds., Cambridge University Press, Cambridge.

Gosline, J. M. (1987) *Rubber Chem. Technol.*, **60**, 417.

Gottlieb, M., Macosko, C. W., Benjamin, G. S., Meyers, K. O., and Merrill, E. W. (1981) *Macromolecules*, **14**, 1039.

Graessley, W. W. (1975a) *Macromolecules*, **8**, 186.

Graessley, W. W. (1975b) *Macromolecules*, **8**, 865.

Greene, A., Smith, K. J., Jr., and Ciferri, A. (1965) *Trans. Faraday Soc.*, **61**, 2772.

Guth, E. and James, H. M., (1941) *Ind. Eng. Chem.*, **33**, 624.

Guth, E. and Mark, H. F. (1934) *Monatsh. Chem.*, **65**, 93.

Herz, J. E., Rempp, P., and Borchard, W. (1978) *Adv. Polym. Sci.*, **26**, 105.

Hinckley, J. A., Han, C. C., Moser, B., and Yu, H. (1978) *Macromolecules*, **11**, 836.

Hoeve, C. A. J. and Flory, P. J. (1974) *Biopolymers*, **13**, 677.

Holden, G. (1973) In *Block and Graft Copolymerization*, Vol. 1, R. J. Ceresa, Ed., Wiley, New York.

Hsu, Y.-H. and Mark, J. E. (1987) *Eur. Polym. J.*, **23**, 829.

Huggins, M. (1942) *J. Phys. Chem.*, **46**, 151.

Ilavsky, M. (1982) *Macromolecules*, **15**, 782.

James, H. M. (1947) *J. Chem. Phys.*, **15**, 651.

James, H. M. and Guth, E. (1947) *J. Chem. Phys.*, **15**, 669.

James, H. M. and Guth E. (1953) *J. Chem. Phys.*, **21**, 1039.

Johnson, R. M. and Mark, J. E. (1972) *Macromolecules*, **5**, 41.

Kilian, H.-G., Enderle, H. F., and Unseld, K. (1986) *Colloid Polym. Sci.*, **264**, 866.

Kuhn, W. (1936) *Kolloid Z.*, **76**, 258.

Kuhn, W. and Grün, F. (1942) *Kolloid-z*, **101**, 248.

Langley, N. R. (1968) *Macromolecules*, **1**, 348.

Leung, Y.-K. and Eichinger, B. E. (1984) *J. Chem. Phys.*, **80**, 3877, 3885.

Liberman, M. H., Abe, Y., and Flory, P. J. (1972) *Macromolecules*, **5**, 550.

Liberman, M. H., DeBolt, L. C., and Flory, P. J. (1974) *J. Polym. Sci., Polym. Phys. Ed.*, **12**, 187.

Llorente, M. A. and Mark, J. E. (1980) *Macromolecules*, **13**, 681.

Llorente, M. A. and Mark, J. E. (1981) *J. Polym. Sci., Polym. Phys. Ed.*, **19**, 1107.

Llorente, M. A., Mark, J. E., and Saiz, E. (1983) *J. Polym. Sci., Polym. Phys. Ed.*, **21**, 1173.

Mark, J. E. (1973) *Rubber Chem. Technol.*, **46**, 593.

Mark, J. E. (1975) *Rubber Chem. Technol.*, **48**, 495.

Mark, J. E. (1976) *Macromol. Rev*, **11**, 135.

Mark, J. E. (1979a) *Polym. Eng. Sci.*, **19**, 254.

Mark, J. E. (1979b) *Polym. Eng. Sci.*, **19**, 409.

Mark, J. E. (1979c) *Acc. Chem. Res.*, **12**, 49.

Mark, J. E. (1979d) *Makromol. Chem., Suppl.*, **2**, 87.

Mark, J. E. (1981) *J. Chem. Educ.*, **58**, 898.

Mark, J. E. (1982a) *Adv. Polym. Sci.*, **44**, 1.

Mark, J. E. (1982b) *Rubber Chem. Technol.*, **55**, 762.

Mark, J. E. (1984a) In *Physical Properties of Polymers*, J. E. Mark, A. Eisenberg, W. W. Graessley, L. Mandelkern, and J. L. Koenig, Eds., American Chemical Society Publications, Washington, DC.

Mark, J. E. (1984b) In *Polymer Chemistry*, American Chemical Society Short Course Manual, J. E. Mark and G. Odian, Eds., American Chemical Society Publications, Washington, DC.

Mark, J. E. (1985a) *Acc. Chem. Res.*, **18**, 202.

Mark, J. E. (1985b) *Br. Polym. J.*, **17**, 144.

Mark, J. E. (1985c) *Polym. J. (Tokyo)*, **17**, 265.

Mark, J. E. (1988) In *Ultrastructure Processing of Ceramics, Glasses, and Composites*, J. D. Mackenzie, and D. R. Ulrich, Eds., Wiley, New York.

Mark, J. E. and Andrady, A. L. (1981) *Rubber Chem. Technol.*, **54**, 366.

Mark, J. E. and Curro, J. G. (1983) *J Chem. Phys.*, **79**, 5705.

Mark, J. E. and Llorente, M. A. (1981) *Polym. J. (Tokyo)* **13**, 543.

Mark, J. E. and Ning, Y.-P. (1985) *Polym. Eng. Sci.*, **25**, 824.

Mark, J. E. and Sullivan, J. L. (1977) *J. Chem. Phys.*, **66**, 1006.

Mark, J. E. and Sung, P.-H. (1982) *Rubber Chem. Technol.*, **55**, 1464.

Mark, J. E. and Yu, C. U. (1977) *J. Polym. Sci., Polym. Phys. Ed.*, **15**, 371.

Mark, J. E., Kato, M., and Ko, J. H. (1976) *J. Polym. Sci., Polym. Symp.*, **54**, 217.

Mark, J. E., Rahalkar, R. R., and Sullivan, J. L. (1979) *J. Chem. Phys.*, **70**, 1794.

Mark, J. E., Ning, Y.-P. Jiang, C.-Y., Tang, M.-Y., and Roth, W. C. (1985) *Polymer*, **26**, 2069.

Mason, P. (1979) *Cauchu. The Weeping Wood*, Australian Broadcasting Commission, Sydney.

Meier, D. J. (1974) *Appl. Polym. Symp*, **24**, 67.

Miller, D. R. and Macosko, C. W. (1987) *J. Polym. Sci. Polym. Phys. Ed.*, **25**, 2441.

Monnerie, L. (1983) *Faraday Symp. Chem. Soc.*, **18**, 57.

Mooney, M. (1948) *J. Appl. Phys.*, **19**, 434.

Morawetz, H. (1985) *Polymers: The Origins and Growth of a Science*, Wiley-Interscience, New York.

Morton, M. Ed. (1973) *Rubber Technology*, Van Nostrand Reinhold, New York.

Mullins, L. and Tobin, N. R. (1965) *J. Appl. Polym. Sci.*, **9**, 2993.

Nagai, K. (1964) *J. Chem. Phys.*, **40**, 2818.

Ning, Y.-P. and Mark, J. E. (1984) *Polym. Bulletin*, **12**, 407.

Noshay, A. and McGrath, J. E. (1977) *Block Copolymers: Overview and Critical Survey*, Academic Press, New York.

Obata, Y., Kawabata, S., and Kawai, H. (1970) *J. Polym. Sci.*, *Part A-2*, **8**, 903.

Ogden, R. W. (1986) *Rubber Chem. Technol.*, **59**, 361.

Pak, H. and Flory, P. J. (1979) *J. Polym. Sci.*, *Polym. Phys. Ed.*, **17**, 1845.

Pearson, D. S. (1977) *Macromolecules*, **10**, 696.

Queslel, J. P. and Mark, J. E. (1984) *Adv. Polym. Sci.*, **65**, 135.

Queslel, J. P. and Mark, J. E. (1985a) *Adv. Polym. Sci.*, **71**, 229.

Queslel, J. P. and Mark, J. E. (1985b) *J. Chem. Phys.*, **82**, 3449.

Queslel, J. P. and Mark, J. E. (1988) In *Comprehensive Polymer Science*, G. Allen, Ed., Pergamon Press, Oxford.

Queslel, J. P., Erman, B., and Monnerie, L. (1985) *Macromolecules*, **18**, 1991.

Read, B. E. and Stein, R. S. (1968) *Macromolecules*, **1**, 116.

Rigbi, Z. (1980) *Adv. Polym. Sci.*, **36**, 21.

Rigbi, Z. and Mark, J. E. (1985) *J. Polym. Sci.*, *Polym. Phys. Ed.*, **23**, 1267.

Rigbi, Z. and Mark, J. E. (1986) *J. Polym. Sci.*, *Polym. Phys. Ed.*, **24**, 443.

Rivlin, R. S. (1948) *Philos. Trans. R. Soc. London, A*, **241**, 379.

Rivlin, R. S. and Saunders, D. W. (1951) *Philos. Trans. R. Soc. London, A.*, **243**, 251.

Ronca, G. and Allegra, G. (1975) *J. Chem. Phys.*, **63**, 4990.

Shen, M., and Croucher, M. (1975) *J. Macromol. Sci.*, *Rev. Macromol. Chem.*, C12, 287.

Smith, K. J., Jr. (1976) *Polym. Eng. Sci.*, **16**, 168.

Smith, T. L. (1977) *Polym. Eng. Sci.*, **17**, 129.

Sohoni, G. B. and Mark, J. E. (1987) *J. Appl. Polym. Sci.*, **33,** 2853.

Sperling, L. H. (1981) *Interpenetrating Polymer Networks and Related Materials*, Plenum Press, New York.

Stein, R. S. (1976) *Rubber Chem. Technol.*, **49,** 458.

Stephens, H. L. (1973) In *Rubber Technology*, M. Morton, Ed., 2nd ed., Van Nostrand Reinhold, New York.

Stepto, R. F. T. (1986) In *Advances in Elastomers and Rubber Elasticity*, J. Lal and J. E. Mark, Eds., Plenum Press, New York.

Su, T.-K. and Mark, J. E. (1977) *Macromolecules*, **10,** 120.

Sun, C.-C. and Mark, J. E. (1987) *J. Polym. Sci., Polym. Phys. Ed.*, **25,** 2073.

Sur, G. S. and Mark, J. E. (1985) *Eur. Polym. J.*, **21,** 1051.

Tanaka, T. (1978) *Phys. Rev. Lett.*, **40,** 820.

Tanaka, T. (1981) *Sci. Am.*, **244**(1), 124.

Treloar, L. R. G. (1975) *The Physics of Rubber Elasticity*, 3rd ed., Clarendon Press, Oxford.

Ullman, R. (1979) *J. Chem. Phys.*, **71,** 436.

Ullman, R. (1982) In *Elastomers and Rubber Elasticity*, J. E. Mark, and J. Lal, Eds., American Chemical Society Publications, Washington, DC.

Urry, D. W., Trapane, T. L., Long, M. M., and Prasad, K. U. (1983) *J. Chem. Soc., Faraday Trans. 1*, **79,** 853.

Volkenstein, M. V. (1963) *Configurational Statistics of Polymeric Chains*, Wiley-Interscience, New York.

Wall, F. T. (1942) *J. Chem. Phys.*, **10,** 485.

Wall, F. T. and Flory, P. J. (1951) *J. Chem. Phys.*, **19,** 1435.

Wang, S.-B. and Mark, J. E. (1987) *Polym. Bulletin*, **17,** 271.

Ward, I. M. (1975) *Structure and Properties of Oriented Polymers*, Applied Science Publishers, London.

Warrick, E. L., Pierce, O. R., Polmanteer, K. E., and Saam, J. C. (1979) *Rubber Chem. Technol.*, **52,** 437.

Weis-Fogh, T. and Andersen, S. O. (1970) *Nature*, **227,** 718.

Yeh, H. C., Eichinger, B. E., and Andersen, N. H. (1982) *J. Polym. Sci., Polym. Phys. Ed.*, **20,** 2575.

Yen, L. Y. and Eichinger, B. E. (1978) *J. Polym. Sci., Polym. Phys. Ed.*, **16,** 121.

Yu, C. U. and Mark, J. E. (1974) *Macromolecules*, **7,** 229.

Zapp, R. L. and Hous, P. (1973) In *Rubber Technology*, M. Morton, Ed., 2nd ed., Van Nostrand Reinhold, New York.

Zhang, Z.-M. and Mark, J. E. (1982) *J. Polym. Sci., Polym. Phys. Ed.*, **20,** 473.

Ziabicki, A. (1976) *Colloid Polym. Sci*, **254,** 1.

Zrinyi, M. and Horkay, F. (1982) *J. Polym. Sci., Polym. Phys. Ed.*, **20,** 815.

BIBLIOGRAPHY

A. Flory, P. J. (1953) *Principles of Polymer Chemistry*, Cornell University Press, Ithaca, New York.

One of the most authoritative texts on polymers, with extensive coverage of rubberlike elasticity. A classic still considered the bible of polymer science in spite of its age. Rigorous, with derivations, and much physical insight.

B. Meares, P. (1965) *Polymers: Structure and Bulk Properties*, Van Nostrand Reinhold, New York.

Good coverage of structure–property relationships of polymers in the bulk state. Particularly good for mechanical properties, including rubberlike elasticity. Some derivations.

C. Dušek, K. and Prins, W. (1969) *Adv. Polym. Sci.*, **6,** 1.

A journal article that is really a short book on the structure of networks and rubberlike elasticity. Quite comprehensive.

D. Chompff, A. J. and Newman, S., Eds. (1971) *Polymer Networks: Structure and Mechanical Properties*, Plenum, New York.

Proceedings of an ACS Symposium on "Highly Cross-Linked Polymer Networks," and thus covers a wide variety of topics.

E. Smith, K. J., Jr. (1972) In *Polymer Science, A. D. Jenkins, Ed., North-Holland, Amsterdam.*

Book chapter on rubberlike elasticity; very useful.

F. Morton, M., Ed. (1973) *Rubber Technology*, Van Nostrand Reinhold, New York.

Several general chapters on applied subjects, followed by more detailed chapters on 20 or so important commercial elastomers.

G. Wall, F. T. (1974) *Chemical Thermodynamics*, 3rd ed., Freeman, San Francisco.
The chapter on "Statistical Thermodynamics of Rubber" very good from a pedagogic point of view.

H. Dunn, A. S., Ed. (1974) *Rubber and Rubber Elasticity*, Polymer Symposium 48, Wiley-Interscience, New York.
Proceedings of a Manchester Symposium honoring L. R. G. Treloar. Twelve articles on various aspects of rubber elasticity.

I. Treloar, L. R. G. (1975) *The Physics of Rubber Elasticity*, 3rd ed., Clarendon Press, Oxford.
The standard book on rubberlike elasticity for the last several decades. Earlier editions have interesting information on crystallization in elastomers.

J. Labana, S. S., Ed. (1977) *Chemistry and Properties of Crosslinked Polymers*, Academic Press, New York.
Proceedings of an ACS symposium having this title, a lot of it on relatively nonelastomeric thermosets.

K. Eirich, F. R., Ed. (1978) *Science and Technology of Rubber*, Academic Press, New York.
Fourteen chapters ranging from polymerization and vulcanization, through rubberlike elasticity and dynamic mechanical properties, to the manufacture of tires.

L. Brydson, J. A. (1978) *Rubber Chemistry*, Applied Science Publishers, London.
Very broad survey as in Eirich (1978) but has much more information on specific elastomers.

M. Nash, L. K. (1979) *J. Chem. Educ.*, **56**, 363.
Very detailed, step-by-step derivation of an elastic equation of state, using the simplest possible approach.

N. Mark, J. E. (1981) *J. Chem. Educ.*, **58**, 898.
Semiquantitative discussions of basic topics, plus some current research areas as well. Strong molecular emphasis.

O. Mark, J. E. and Lal, J., Eds. (1982) *Elastomers and Rubber Elasticity*, American Chemical Society Publications, Washington.
Proceedings of an ACS symposium having this title. About a third involves the preparation of elastomers and the rest their characterization.

P. Eichinger, B. E. (1983) *Annu. Rev. Phys. Chem.*, **34**, 359.
Thermodynamics and continuum mechanics used to cover the subject of "The Theory of High Elasticity." Very useful comments on unsolved problems, particularly in the area of theory.

Q. Labana, S. S. and Dickie, R. A., Eds. (1984) *Characterization of Highly Cross-Linked Polymers*, American Chemical Society Publications, Washington, DC.
Proceedings of an ACS symposium that is a sequel to Labana (1977).

R. Lal, J. and Mark, J. E., Eds. (1986) *Advances in Elastomers and Rubber Elasticity*, Plenum, New York.

Proceedings of an ACS symposium that is a sequel to Mark and Lal (1982).

S. Queslel, J. P. and Mark, J. E. (1986) in *Wiley-Interscience Encyclopedia of Polymer Science and Engineering*, 2nd ed., Wiley-Interscience, New York. (Article title is "Elasticity.") Coverage of a variety of topics, including ASTM test methods.

T. Singler, R. E. and Byrne, C. A., Eds. (1987) *Elastomers and Rubber Technology*, U.S. Government Printing Office, Washington, DC.
Proceedings of the Sagamore Army Materials Research conference having this title. Particularly good for more applied topics.

U. Queslel, J. P. and Mark, J. E. (1987) in *Encyclopedia of Physical Science and Technology*, R. A. Meyers, Ed., Academic Press, New York (Article title is "Rubberlike Elasticity.")
Extensive survey, with a lot of information on chain statistics.

V. Queslel, J. P. and Mark, J. E. (1987) *J. Chem. Educ.*, **64,** 491.
Pedagogic treatment that is a sequel to Mark (1981).

AUTHOR INDEX

187

SUBJECT INDEX